W0007154

100% PROOF !

THE WHY OF MATHS

Visual and algebraic explanations of formulas

needed for GCSE and A level Mathematics

JANE HOPE

© 2015 Jane Hope

All rights reserved.

ISBN 978-0-9935722-2-7

Printed in United Kingdom

First printing 2015

www.pythag.co.uk

Preface

This book is a collection of over a hundred visual and algebraic explanations and proofs of maths formulas and concepts, all needed for GCSE and A level. Set out as very understandable individual pages per proof, with clear diagrams and simple algebra, all these formulas are found within one handy volume.

Included are :
Areas of rectangle, circle, triangle, parallelogram, trapezium, surface area and volume cone and sphere, ratio sides of triangle, Pythagoras, sine relationship, sine formula, sine area triangle, cosine rule, quadratic formula to find root, root relationships, circle theorems, trig identities, double angle, sums of sines, cosines, difference of two squares, completing the square, sum of geometric and arithmetic sequences, formula for circle.........among others

Often the 'why of maths' gets overlooked for the 'how' of using it, but these simple visual pages act as a key to understanding, and once understood, the maths becomes easier, it makes more sense, and the formulas are more likely to be remembered.

The skills developed by using these pages are also fundamental to problem solving, finding an unknown, building on what you know to be true, experimenting, using logic. Such skills are essential for exam questions; problem solving is an important feature in current maths courses, as well as being a useful life skill.

This book is a good revision addition, key to understanding and memory aid.

Introduction

All too often we have to learn the HOW of maths, focussing on how to use the concepts and formulas, without understanding why they exist. This book explains the WHY of maths that we use at school, or maybe left school without fully understanding. It explains as simply as possible all those well known formulas with the use of visuals and some algebra, so that we can SEE the truth of the maths.

Throughout history maths theorems and formulas have been created or discovered, but they have all been derived from simple truths, many building on each other. This book shows that we use 'things we know to be true' to prove other facts or theorems. Some examples of these simple truths are : angles on a straight line summing to 180 degrees, the parallel line properties which make intercepted angles equal or radii of a circle being equal .

Any triangle can be split into two right angled triangles by dropping a

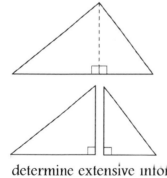

perpendicular. This is a most useful tool ; the resulting facts that we can ascertain eg. length of sides (Pythagoras) or of angle size (trigonometry from similar triangles) coupled with the notion that any straight sided polygon can be converted into a number of triangles shows that we can determine extensive information about any straight sided shape. Often in proof we need to add lines, these can be parallel or perpendicular, or just to divide shapes, or realise certain facts, such as the two acute angles in a right angle triangle sum to 90 degrees.

For instance if you drop a perpendicular from the top, between the two equal lines of an isosceles triangle you will create two congruent triangles, (2 sides and a right angle equal), therefore the two base angles are equal and the base is bisected. This can then be applied to any triangle within a circle made by two equal radii and connecting chord.

It is a creative process, and sometimes one has to think like a detective, 'What do we have?' 'What can we do?' and let the process take you on a journey.

This is also an important part of problem solving, knowing what to do when you don't know the answer. This skill needs to be practised since it is essential for the type of problem solving needed in any examination, when we have to make sense of unfamiliar questions.

Visuals are a very powerful tool; we have a strong visual side to our brains. Maths is in itself a feat of hand, eye, brain coordination. Very good exam advice is – 'if in doubt draw a diagram'. Visuals can be used for many concepts, not just shape, although we do in fact live on a sphere, which revolves in a circular orbit about another sphere, our sun. In fact, due to the existence of gravity, the sphere must be the most common shape in our galactic universe. I find it fascinating that Archimedes found that the surface area of a sphere is equal to the surface area of an open cylinder. It is true that the great mathematical scientist of quantum theory Paul Dirac was happiest when he could visualise difficult concepts.

As shown in the book, numbers can be represented in patterns or as rectangles, which is a very useful tool in ratio. In fact Singapore Maths uses the rectangle and visuals for much of its problem solving work, which is an important part of its system. Perhaps this is part of the reason their students do so well. Internationally they are at the top of the rankings, while we are much lower in comparison.

v

Strategies for problem solving (also used by Singapore) include: use a diagram, look for a pattern, simplify problem, work backwards, use algebra, look for clues, experiment and finally reflect. This book encourages the reader to do all these things, by looking at the diagrams, finding the clues, and ultimately finding the truth of the maths for themselves. I hope you will be encouraged to experiment for yourself, and see where an initial idea may lead.

During my own study I was blown away when I first saw the original Pythagoras proof by its beautiful simplicity, just by re-arrangeing four congruent right angled triangles within a square (p71), and linking to the algebra in a simple, elegant way.   At the time I thought 'Wouldn't it be great if there was a book that did this for all the theorems we use', but couldn't find one – so started writing it myself.   In my teaching I have used it extensively, and in my long experience as a personal maths tutor I am very aware of the most common mistakes and misunderstandings, so have tried to include all the things people find most tricky.

The Pythagorean Theorem is possibly the most famous in the history of mathematics, with many hundred proofs. I have included a whole section of eight colourful proofs, which I felt were ideal to use as a classroom resource, and had made as laser cut acrylic puzzles– and used with success. Although known as proofs by dissection, I have included my own proofs for these, with angle facts, parallel lines, which show the truth of the theorem with minimal algebra.

100% PROOF !

THE WHY OF MATHS

GCSE content

100% PROOF !

THE WHY OF MATHS

A Level contents

BASICS

Angle Sum of a Triangle

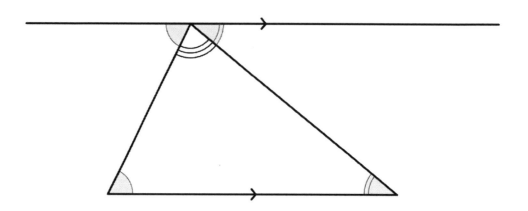

Things we know to be true

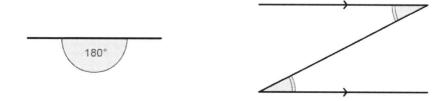

Angles on a straight line sum to 180°   Alternate angles of parallel lines are

equal

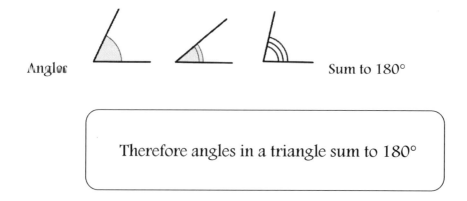

Angles   Sum to 180°

Therefore angles in a triangle sum to 180°

Exterior Angle of a Triangle

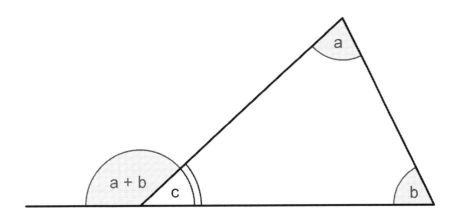

Angles $a + b + c = 180°$

(Angles of a triangle or a straight line equal 180°)

Therefore exterior angle of a triangle equals

the sum of the 2 opposite angles

Parallel Lines

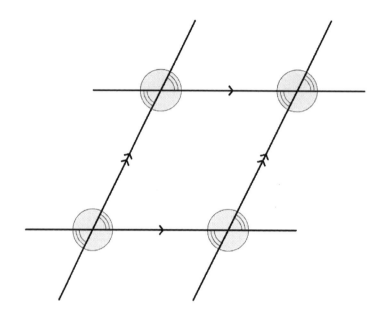

Opposite angles are equal

Corresponding (F) angles are equal

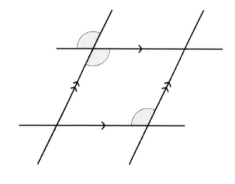

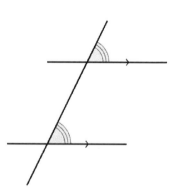

Alternate (Z) angles are equal

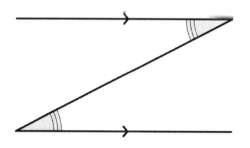

Fractions : Add Subtract Multiply Divide

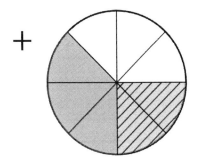

+

$$\frac{3}{8} + \frac{1}{4} = \qquad \frac{3+2}{8} = \frac{5}{8}$$

Fractions must have the same denominator to be

able to add them $\qquad \dfrac{numerator}{denominator}$

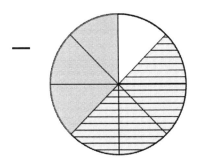

−

$$\frac{7}{8} - \frac{1}{2} = \qquad \frac{7-4}{8} = \frac{3}{8}$$

or subtract them

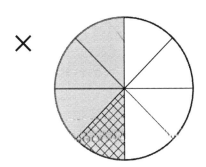

×

$$\frac{1}{4} \times \frac{1}{2} = \qquad \frac{1}{8}$$

$$think \, \frac{1}{4} \, of \, \frac{1}{2}$$

Multiply numerators and also denominators

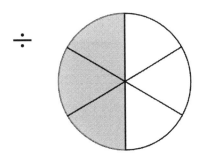

÷

$$\frac{1}{2} \div \frac{1}{6} = \qquad 3$$

$$\equiv \frac{1}{2} \times \frac{6}{1} = 3$$

Think how many sixths there are in one half

Flip 2nd fraction and turn into multiply

Positive and Negative Numbers

The number 5 ...... and taking away a negative number

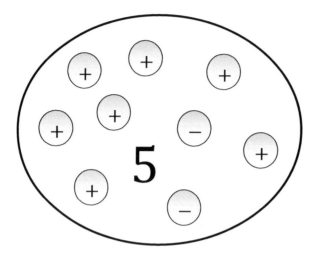

The number 5 can be thought of as $5 + 2 - 2$

.... the plus and minus symbols cancel each other out

To subtract a negative number

$$5 - -2 = ?$$

Take away the two $-ve$ symbols

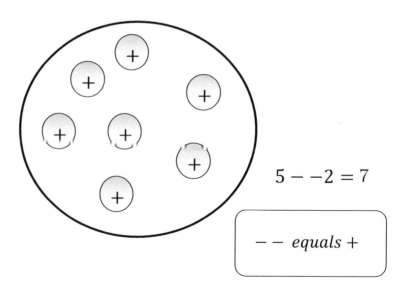

$$5 - -2 = 7$$

$-- \; equals \; +$

Multiplying Positive and Negative Numbers

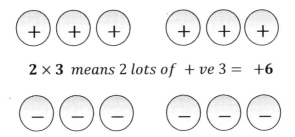

2 × 3 *means 2 lots of + ve 3 =* **+6**

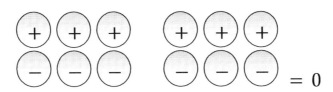

2 × −3 *means 2 lots of − ve 3 =* **−6**

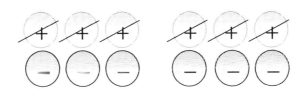

Zero is made of an equal number of + ve and − ve

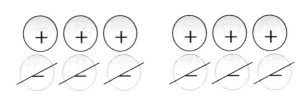

−2 × 3 *means take away 2 lots of + ve 3* = **−6**

−2 × −3 *means take away 2 lots of − ve 3* = **+6**

Proportion and Ratio

Any amount can be represented by a rectangle

An amount **P** is divided into ratio 4 : 3 : 2 : 1

First add all the shares together,   then divide the amount P by this total – to get a value for **one share – or one box**

$$4 + 3 + 2 + 1 = 10$$

One share $= \dfrac{1}{10}$ **of P** $= \dfrac{P}{10}$

Then it is easy to find the value of 4 yellow shares,

or 3 blue etc

Nth Term

a = first term   d = difference   n = term number

each term is the first term plus the difference times one less than n

$$nth\ term =\ a + (n-1)d$$

OR

Zero
term

Find the zero term − each term is the difference times n added to this

$$nth\ term =\ difference \times n + zero\ term$$

Algebra Rules

Algebra means using **letters** or **symbols** for numbers to find **unknown** quantities.   A **variable** is a letter whose number value can vary

~~~~~~

$$x + x + x = 3 \times x = \mathbf{3x}$$

There is an invisible × sign between the 3 and the $x = 3x$

There is also an invisible × sign next to brackets

$$\overset{\times}{3^{\downarrow}(x+1)} \qquad \overset{\times}{(x+3)^{\downarrow}(x-2)}$$

~~~~~~

Letters can be added or subtracted only if they are of the same type

$$eg\ 3x + 2x = 5x$$

$$x^2 = x \times x \qquad x^2\ is\ an\ area,\ x^3\ is\ a\ volume$$

and cannot be added to x except as a polynomial

$$x^3 + x^2 + x$$

x

length

x^2

square

x^3

cube

Brackets and Multiplying

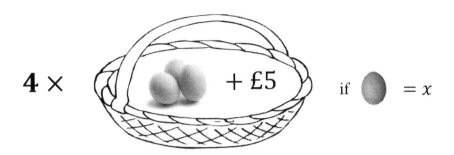

Think of brackets as baskets – so 4 baskets means 4 times everything in the basket

$$4(3x + 5) \quad = \quad 4 \times 3x \quad + \quad 4 \times 5$$

$$= 12x + 20$$

or 12 eggs and £20

~~~~~~

Two brackets $\quad (x + 3)(x + 2) \quad$ are the same as

x ⬡ $x + 2$ $\quad + \quad 3$ ⬡ $x + 2$

$$x(x + 2) \quad + \quad 3(x + 2)$$

$$x^2 + 2x \quad + \quad 3x + 6 =$$

$$= x^2 + 5x + 6$$

11

Algebra – Balancing Equations

$$3x + 2 = x + 6$$

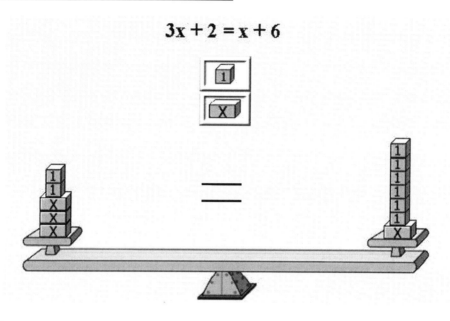

Think of an equation as a weigh balance – either side of the equal sign –

If you do the same to both sides they remain equal

(*do to both sides*)	$3x + 2 = x + 6$
$(-x)$	$2x + 2 = 6$
(-2)	$2x = 4$
$(\div 2)$	$x = 2$

~~~~~~

Rules – do the **same** to both **sides**

Do the **opposite** operation until you have isolated x one side of the equal sign

$$+ 2 \quad \leftrightarrow \quad - 2$$

$$\times 3 \quad \leftrightarrow \quad \div 3$$

$$\blacksquare^2 \quad \leftrightarrow \quad \sqrt{\blacksquare}$$

(Multiply out any brackets first), then rearrange

ANGLES

Regular Polygon Interior Angle

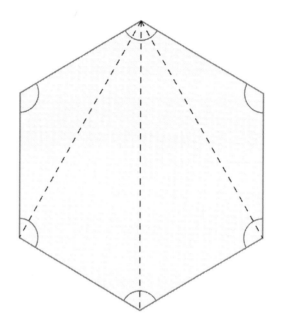

A polygon of n sides can be divided into $n - 2$ triangles

Therefore the interior angles sum to $180° \times (n - 2)$

$$\therefore interior\ angle - \frac{180}{n}(n - 2)$$

Regular Polygon Exterior Angle

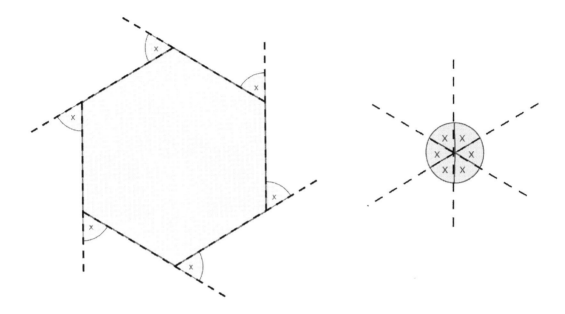

The exterior angles of a regular polygon are all equal, and sum to $360°$

If n is number of sides of polygon

$$\boxed{exterior\ angle\ =\ \frac{360°}{n}}$$

$$\therefore\ interior\ angle\ =\ 180°\ -\ \frac{360°}{n}$$

$$=\ 180\left(1-\frac{2}{n}\right)\ =\ \frac{180}{n}(n-2)$$

Similar Triangles

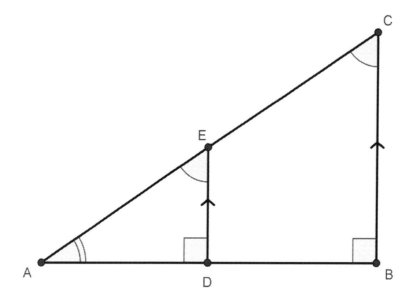

If two triangles have the same angles, there will be a scale factor S which will make one triangle congruent to the other

These triangles will be in ratio with each other

> ## Corresponding sides will be in ratio

$$\frac{AD}{AB} = \frac{AE}{AC} = \frac{DE}{BC}$$

$$AD \times \textbf{\textit{S}} = AB \quad \therefore \quad \textbf{\textit{S}} = \frac{AB}{AD}$$

Sine, Cosine & Tangent from Ratio Triangles

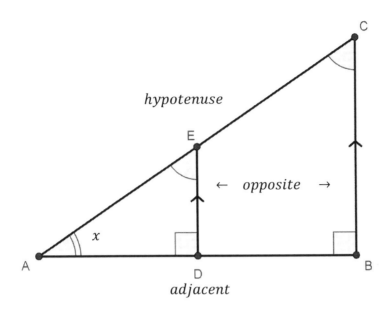

Constant ratios of sides of right angled triangles in relation to angles are used in trigonometry

$$eg.\,For\ angle\ x, \qquad ratios\ \frac{DE}{AE} = \frac{BC}{AC} \qquad and\ are\ called\ Sin\ x$$

$$Sin\ x = \frac{opposite}{hypotenuse} = \frac{O}{H}$$

Similarly

$$Cos\ x = \frac{adjacent}{hypotenuse} = \frac{A}{H}$$

And

$$Tan\ x = \frac{opposite}{adjacent} = \frac{O}{A}$$

SOH CAH TOA Some Old Horse Caught Another Horse Taking Oats Away

Scale Factors

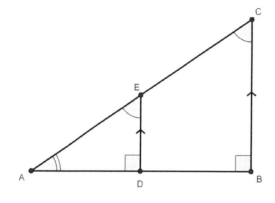

If lines are in ratio as in similar triangles with scale factor S

$$Then\ AD \times S = AB$$

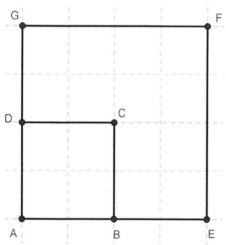

$$AB \times S = AE \quad and \quad AD \times S = AG$$

$$\therefore \ AE \times AG = AB \times AD \times S^2$$

area large square
$$= \text{area small square} \times S^2$$

Therefore for area the scale factor is squared

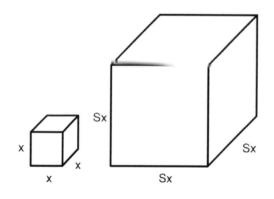

Scale factor S makes volume

$$x^3\ become\ S^3 x^3$$

Therefore for volume the

scale factor is cubed

AREA & VOLUME

Area Rectangle

Area Circle

Length of Arc Circle and Area of Sector

Area of Triangle

Area of Parallelogram

Area of Trapezium

Surface Area of Cone

Surface Area of Sphere

Volume of Prism

Volume of Cone or Pyramid

Volume of Sphere

Area Rectangle

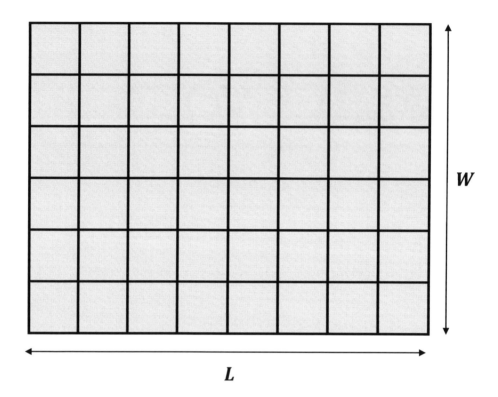

L boxes in each row, **W** amount of rows

$$Area = L \times W$$

Boxes = 1 unit square.

$$Area \ = \ Length \ \times \ Width$$

Area Circle

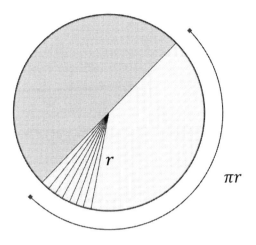

If you cut a circle into very thin segments and rearrange, a rectangle of length πr (half the circumference of $2\pi r$) by r (radius) is formed.

Area of rectangle $=$ $length \times width$ $=$ $\pi r \times r$ $=$ πr^2

$$Area\ \ Circle\ =\ \pi r^2$$

Length of Arc Circle and Area of Sector

Fact := *Perimeter of circle is* $\pi \times diameter = 2\pi r$

<u>Definition of π</u> – π is how many times the diameter of a circle goes around the circumference (a little over 3 times) (3.14159 26535 89793 23846 26433 83279 50288 41971 69399 37510 58209 74944 59230 78164 06286 20899 86280 34825 34211 70679 82148 08651 ....)

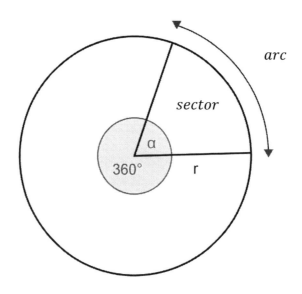

Fraction of circle which gives sector or arc with angle α, is given by $\dfrac{\alpha}{360}$

$$Arc\ length = \frac{a}{360} \times 2\pi r$$

$$Area\ sector = \frac{a}{360} \times \pi r^2$$

Area of Triangle

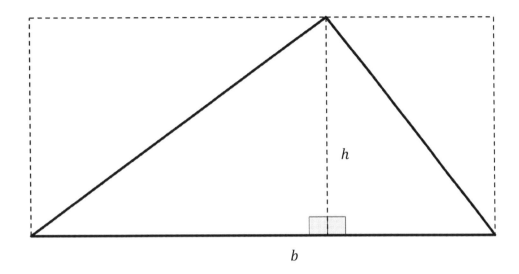

Drop a perpendicular to base of triangle. Each right angled triangle formed is half a rectangle, which when added together make rectangle of base b and height h

Area of original triangle is therefore half of this rectangle

$$Area\ triangle = \frac{1}{2}\ base\ \times perpendicular\ height$$

$$\boxed{Area\ triangle\ =\ \frac{1}{2}\ b\ \times h}$$

Area of Parallelogram

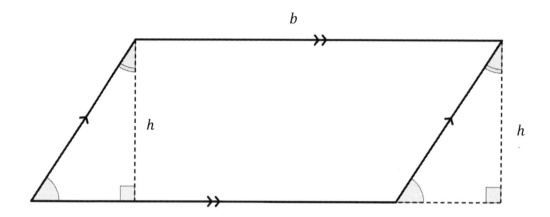

The triangles formed when you drop perpendicular lines from the top two corners to the base line are congruent (equal angle properties of parallel lines) hence a rectangle is formed, base b, height h, with same area as the parallelogram,

Area rectangle = length × width = base (b) × height (h) (perpendicular)

Area parallelogram = base × perpendicular height

Area parallelogram = $b \times h$

Area of Trapezium

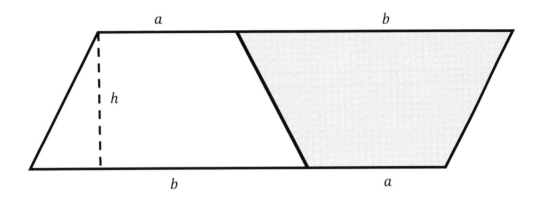

Take 2 identical trapeziums, one inverted and join,

the parallelogram formed has $Area = (a + b) \times h$

where h is perpendicular height.

Therefore one trapezium has area equal to half the parallelogram

$Area =$ **half the sum of parallel sides** $\times$ **perpendicular height**

$$Area\ trapezium = \frac{1}{2}(a+b)h$$

Surface Area of Cone

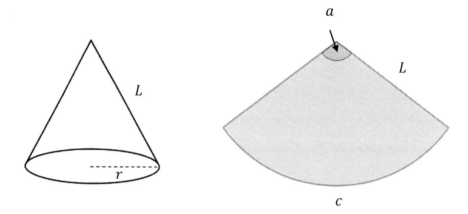

The net of a hollow cone is a sector of circle with radius L

Area and arc length of sector calculated as fraction of whole circle

$$Area\ sector\ of\ circle\ =\ \frac{a}{360} \times \pi L^2 \qquad eq(i)$$

Arc length c   =   circumference of the base of the cone

$$\frac{a}{360} \times 2\pi L\ =\ 2\pi r \qquad (\div\ 2\pi\ both\ sides)$$

$$\therefore \frac{a}{360} \times L\ =\ r \qquad (substitute\ in\ eq\ (i))$$

$$Area\ sector\ circle = \left(\frac{a}{360} \times L\right) \times \pi L\ =\ \pi r L$$

$$Surface\ area\ cone\ =\ \pi r L$$

Surface Area of Sphere

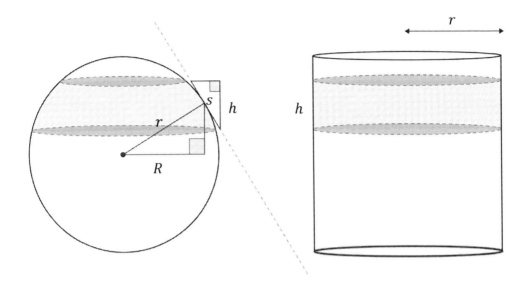

Take a slice of sphere and corresponding cylinder, height **h**. Slope of sphere modelled here by tangent length **s**. 2 congruent angled triangles are formed as shown.

$$Surface\ area\ of\ cylinder\ slice = 2\pi rh$$

$$Surface\ area\ of\ sphere\ modelled\ slice := 2\pi Rs$$

$$By\ similar\ triangles\ \frac{R}{r} = \frac{h}{s}$$

$$\therefore\ Rs = rh$$

$$\therefore\ 2\pi Rs = 2\pi rh\ \ and\ area\ of\ slices\ the\ same$$

As number of slices tends to infinity the area of whole cylinder equals the area of sphere. Therefore

$$\textbf{\textit{Surface area of sphere = area of open cylinder}}$$
$$\textbf{\textit{= } } 4\pi r^2$$

Volume of Prism

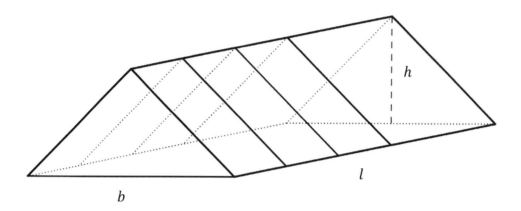

A prism can be cut into equal slices of unit length

Therefore the volume is the **area of the cross section** times the **length**

This is true for any 3D shape for which the cross section is unaltered throughout its length eg. cylinder or cuboid

Volume triangular prism $= \dfrac{1}{2} base \times height \times length$

Volume cuboid $= b \times h \times l$

Volume cylinder $= \pi r^2 h$

Volume of Cone or Pyramid

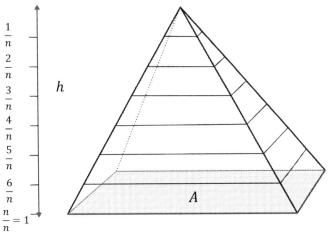

scale factor

$\frac{1}{n}$
$\frac{2}{n}$
$\frac{3}{n}$
$\frac{4}{n}$
$\frac{5}{n}$
$\frac{6}{n}$
$\frac{n}{n} = 1$

h

A

Take any sided pyramid, perpendicular height h, and base area A, split into n vertical layers.

From similarity, the area of the base of each cross section will be A times the scale factor squared :

$$from\ top\ A\left(\frac{1}{n}\right)^2 ,A\left(\frac{2}{n}\right)^2 ,\ A\left(\frac{3}{n}\right)^2 ,\ \ ...\ A\left(\frac{n}{n}\right)^2\ at\ base$$

The volume can be approximated (n large) as each of these areas times h / n

$$V := \frac{h}{n}\left(\frac{1}{n}\right)^2 A + \frac{h}{n}\left(\frac{2}{n}\right)^2 A + \frac{h}{n}\left(\frac{3}{n}\right)^2 A + \cdots \frac{h}{n}\left(\frac{n}{n}\right)^2 A$$

$$= \frac{Ah}{n^3}(1^2 + 2^2 + 3^2 + \cdots n^2)$$

*Using **sum of squares*** $1^2 + 2^2 + 3^2 ... n^2 = \frac{1}{6}n(n + 1)(2n + 1)$

$$V = \frac{Ah}{n^3} \times \frac{1}{6}n(n + 1)(2n + 1) = \ \ \frac{Ah}{6n^3}(n^3 + 3n^2 + n) = \ \ \frac{Ah}{6}\left(1 + \frac{3}{n} + \frac{1}{n^2}\right)$$

$$\therefore V = \frac{Ah}{6}\left(1 + \frac{1}{n}\right)\left(2 + \frac{1}{n}\right) \quad As\ n\ \rightarrow\ \infty\ \ \frac{1}{n} \rightarrow 0\ \therefore V = \frac{Ah}{6} \times 2$$

$$\boxed{\textbf{\textit{Therefore}}\ \ \textbf{\textit{V}} = \frac{1}{3}\ \textbf{\textit{Ah}}}$$

Since this can be any sided pyramid, this also holds for cone volume

Volume of Sphere

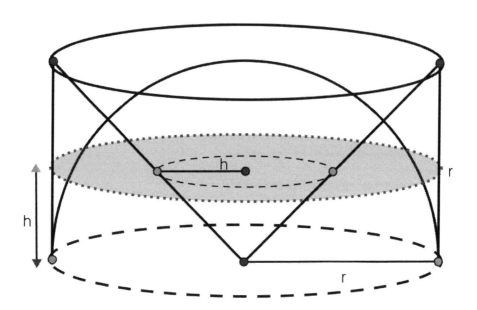

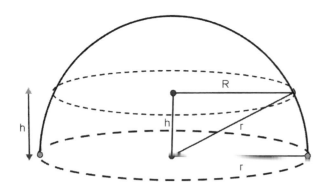

Radius hemisphere slice

$$R = \sqrt{r^2 - h^2}$$

A hemisphere and an inverted cone are inscribed in a cylinder, all having radius r. The point of the cone is at the centre of the circle base.

On a horizontal slice at height h, the area of the cross sections are :

$$cylinder := \pi r^2 \quad cone := \pi h^2 \quad hemisphere := \pi(r^2 - h^2)$$

Area cylinder minus area cone = area hemisphere (at any horizontal slice cross section)

Therefore the volume of the hemisphere at any height (including full height r) will be the difference between the volume of the cylinder and cone.

And the volume of the complete hemisphere height r will be

$$Vol\ hemisphere = \pi r^3 - \frac{1}{3}\pi r^3 = \frac{2}{3}\pi r^3$$

$$Therefore\ \textbf{Volume of sphere} = \frac{4}{3}\pi r^3$$

ALGEBRA

Difference of Two Squares

Completing the Square

Completing the Square(algebra)

Quadratic Roots – from completing the square

Root Relationships

Difference of Two Squares

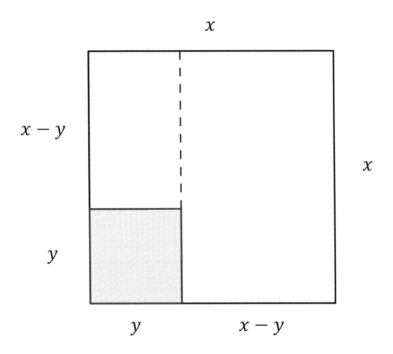

Subtract the area of the small square from the larger, leaving two rectangles

$$x^2 - y^2 = x(x - y) + y(x - y)$$

$$\therefore \quad x^2 - y^2 = (x + y)(x - y)$$

Completing the Square

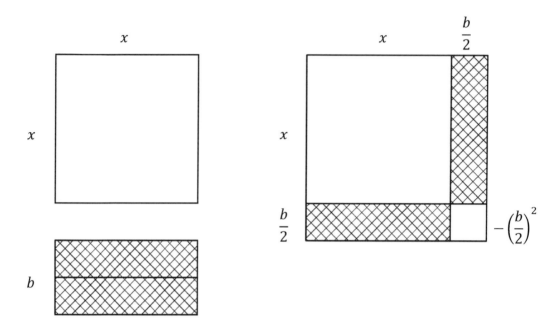

$$x^2 + bx = \left(x + \frac{b}{2}\right)^2 - \left(\frac{b}{2}\right)^2$$

$$x^2 + bx + c = \left(x + \frac{b}{2}\right)^2 - \left(\frac{b}{2}\right)^2 + c$$

Completing the Square(algebra)

$$y \quad = \quad ax^2 + bx + c$$

$$y \quad = \quad a\left(x^2 + \frac{b}{a}x\right) + c \qquad - eq.\,(i)$$

$$\left(x + \frac{b}{2a}\right)^2 \quad = \quad x^2 + \frac{b}{a}x + \left(\frac{b}{2a}\right)^2$$

$$\therefore \quad x^2 + \frac{b}{a}x \quad = \quad \left(x + \frac{b}{2a}\right)^2 - \left(\frac{b}{2a}\right)^2$$

Substitute RHS in above - $eq.\,(i)$

$$y \quad = \quad a\left(\left(x + \frac{b}{2a}\right)^2 - \left(\frac{b}{2a}\right)^2\right) + c$$

$$y \quad = \quad a\left(x + \frac{b}{2a}\right)^2 + \left(c - \frac{b^2}{4a}\right)$$

and if $a = 1$

$$y \quad = \quad \left(x + \frac{b}{2}\right)^2 - \left(\frac{b}{2}\right)^2 + c$$

a *is scaling factor,*

$\dfrac{b}{2a}$ *is displacement of x,*

c *is value of y when x = 0 ,*

$\left(c - \dfrac{b^2}{4a}\right)$ *is value of y at turning point*

Quadratic Roots - from completing the square

$$ax^2 + bx + c = 0$$

$$\therefore a\left(x + \frac{b}{2a}\right)^2 + \left(c - \frac{b^2}{4a}\right) = 0$$

$$a\left(x + \frac{b}{2a}\right)^2 = \frac{b^2}{4a} - c$$

$$\therefore \left(x + \frac{b}{2a}\right)^2 = \frac{b^2 - 4ac}{4a^2}$$

Take square roots both sides

$$\therefore \left(x + \frac{b}{2a}\right) = \frac{\pm\sqrt{b^2 - 4ac}}{2a}$$

$$\therefore x = \frac{\pm\sqrt{b^2 - 4ac}}{2a} - \frac{b}{2a}$$

$$\boxed{\therefore \quad x = \frac{-b \pm \sqrt{b^2 - 4ac}}{2a}}$$

Root Relationships

$$x = \frac{-b + \sqrt{b^2 - 4ac}}{2a} \qquad \text{or} \qquad x = \frac{-b - \sqrt{b^2 - 4ac}}{2a}$$

Sum of Roots

$$\frac{-b + \sqrt{b^2 - 4ac}}{2a} \quad + \quad \frac{-b - \sqrt{b^2 - 4ac}}{2a}$$

$$= -\frac{2b}{2a}$$

$$\boxed{= -\frac{b}{a}}$$

Product of Roots

$$\frac{-b + \sqrt{b^2 - 4ac}}{2a} \quad \times \quad \frac{-b - \sqrt{b^2 - 4ac}}{2a}$$

$$= \frac{b^2 - (b^2 - 4ac)}{4a^2}$$

$$= \frac{4ac}{4a^2}$$

$$\boxed{= \frac{c}{a}}$$

GRAPHS

Equation of a Straight Line

Gradients

Perpendicular Gradient

Midpoint of Straight Line

Distance Between Two Points

Graph Translations – Graph of x^2

Graph Stretch Factors

Graph – solving from completing the square

Formula for Circle

Formula for any Circle

Equation of a Straight Line

Graphs of lines with varying gradient or slope (m)

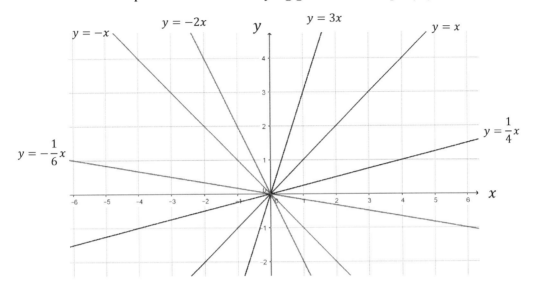

Graphs of lines with varying y intercept (c) but with same slope(m)are parallel

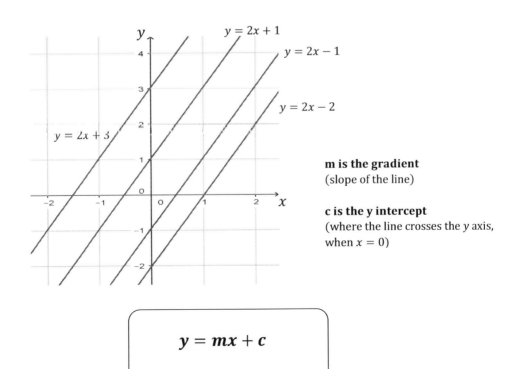

m is the gradient
(slope of the line)

c is the y intercept
(where the line crosses the y axis,
when $x = 0$)

$$y = mx + c$$

Gradients

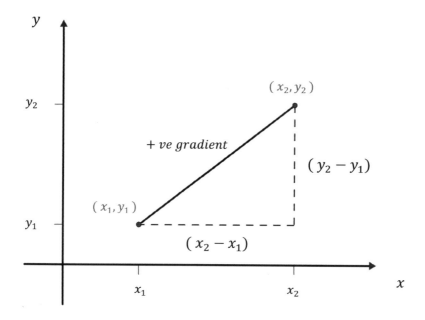

$$\text{Gradient } (m) \quad = \quad \frac{height}{base}$$

$$m \quad = \quad \frac{(y_2 - y_1)}{(x_2 - x_1)}$$

$(x_2 - x_1)$ *must always be* $+ve$ *if* x_1 *on left*

If $y_2 > y_1$ *then gradient is* $+ve$

If $y_2 < y_1$ *then gradient is* $-ve$

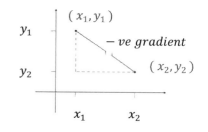

rearrange for equation of line

$$y - y_1 = m(x - x_1)$$

Perpendicular Gradient

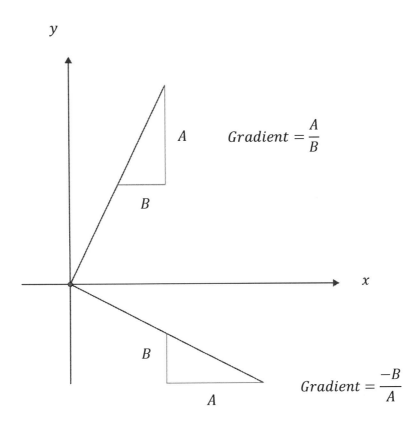

Right angled triangle of sides A & B is rotated through 90°

If Gradient $\dfrac{A}{B} = m$ Gradient $-\dfrac{B}{A} = -\dfrac{1}{m}$

$$\therefore \; m_{perp} = -\dfrac{1}{m}$$

Midpoint of Straight Line

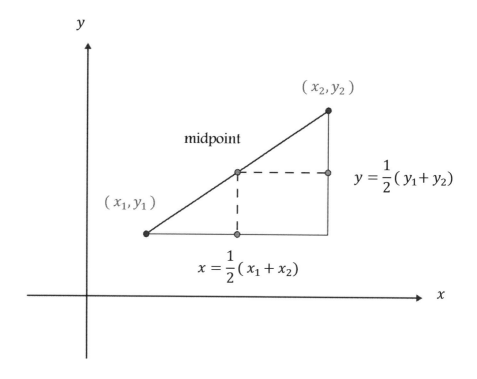

Co-ordinates of the midpoint of a straight line are an average of the x , y values. For points with coordinates (x_1, y_1) and (x_2, y_2)

Midpoint of straight line is

$$\left(\frac{x_1 + x_2}{2} , \frac{y_1 + y_2}{2} \right)$$

Distance Between Two Points

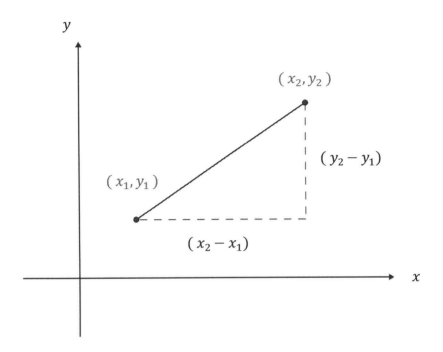

From Pythagoras the distance between two points

with coordinates (x_1, y_1) *and* (x_2, y_2) is

$$\sqrt{(x_2 - x_1)^2 + (y_2 - y_1)^2}$$

Graph Translations – Graph of x^2

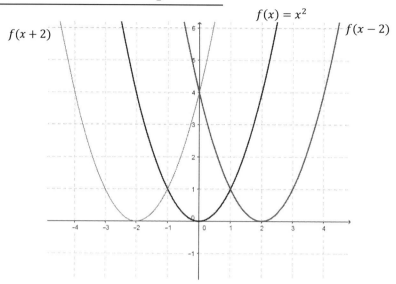

$y = (x - 2)^2$ *translates in opposite x (+ve) direction* $= f(x - 2)$

$y = (x + 2)^2$ *translates in opposite (−ve) x direction* $= f(x + 2)$

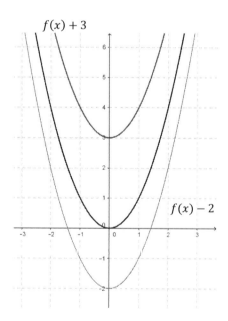

$y = x^2 + 3$ *translates vertically in same + ve direction* $= f(x) + 3$

$y = x^2 - 2$ *translates vertically in same − ve direction* $= f(x) - 2$

Graph Stretch Factors

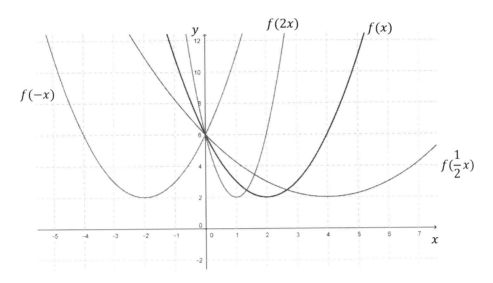

$y = f(ax)$ is a stretch in the x direction scale factor $\dfrac{1}{a}$

$y = f(-x)$ is a reflection in the y axis

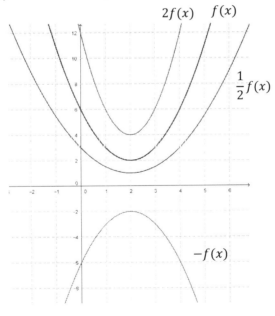

$y = af(x)$ is a stretch in the y direction scale factor a

$y = -f(x)$ is a reflection in the x axis

45

Graph – solving from completing the square

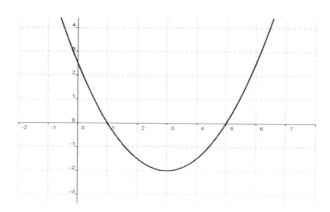

$$y = \frac{1}{2}(x-3)^2 - 2 \quad \text{Minimum point } (3,-2)$$

Roots x = 1 , x = 5 (when y = 0) $y\ intercept\ \left(0, \frac{5}{2}\right)$ $y = \frac{1}{2}x^2 - 3x + \frac{5}{2}$

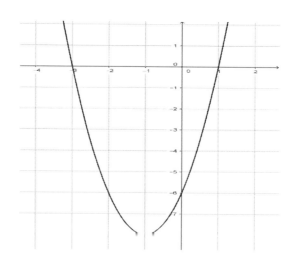

$$y = 2(x+1)^2 - 8 \quad \text{Minimum point } (-1,-8)$$

Roots x = −3, x = 1 (when y = 0) $y\ intercept\ (0,-6)$ $y = 2x^2 + 4x - 6$

$$y = \frac{1}{2}x^2 - 3x + \frac{5}{2} \qquad a = \frac{1}{2} \ \ b = -3 \ \ c = \frac{5}{2}$$

or $y = 2x^2 + 4x - 6 \qquad a = 2 \ \ b = 4 \ \ c = -6$

and use formula completing the square

$$= a\left(x + \frac{b}{2a}\right)^2 + \left(c - \frac{b^2}{4a}\right)$$

a is scaling factor,

$\dfrac{b}{2a}$ is displacement of x,

c is value of y when $x = 0$,

$\left(c - \dfrac{b^2}{4a}\right)$ is value of y at turning point

Formula for Circle

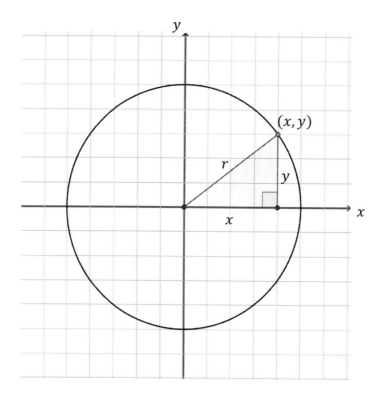

From any point on a circle (x, y), a line to the centre, radius r, and a perpendicular to the x axis, produces a right angled triangle of sides x, y and hypotenuse r

From Pythagoras

$$x^2 + y^2 = r^2$$

Formula for any Circle

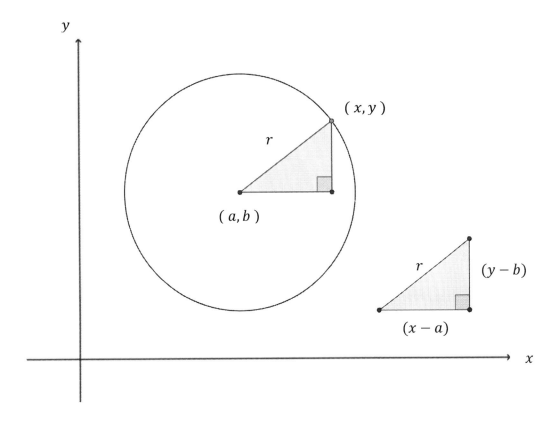

For circle centre (a, b), radius r, a point (x, y) will make a right angle triangle as shown, with sides $(x - a), (y - b)$ and r

From Pythagoras

$$(x - a)^2 + (y - b)^2 = r^2$$

CIRCLE THEOREMS

Double Angle at Centre

Equal Angles at Circumference

Cyclic Quadrilateral

Diameter makes Right Angle

Tangent Properties

Alternate Segment

Double Angle at Centre

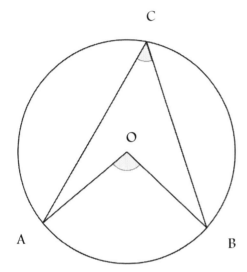
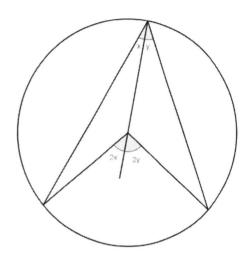

Draw line from C through centre.   Two isosceles triangles are formed,

AOC and BOC.   The exterior angles formed are 2x and 2y .

Hence angle AOB = 2(x+y) = double angle ACB (x+y)

The angle from two points on a circle to the centre

is double the angle to the circumference

Equal Angles at Circumference

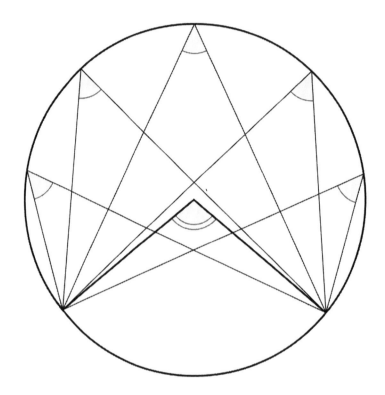

Because all the angles at the circumference are subtended from the same two points, they are all half the angle at the centre, hence they are all equal to each other.

Angles from the same 2 points to the circumference are equal

Cyclic Quadrilateral

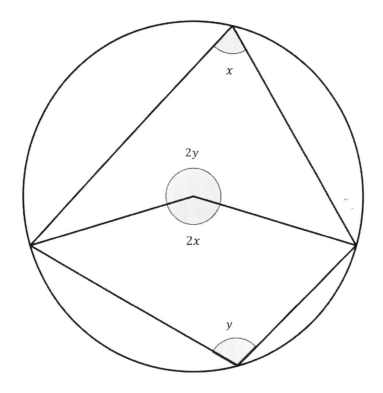

From the double angle theorem, angle marked at centre $2y$ is double y, likewise $2x$ is double x

$$2x + 2y = 360° \ at \ centre$$

$$\therefore \ x + y = 180°$$

Therefore opposite angles of a cyclic quadrilateral sum to $180°$

Diameter makes Right Angle

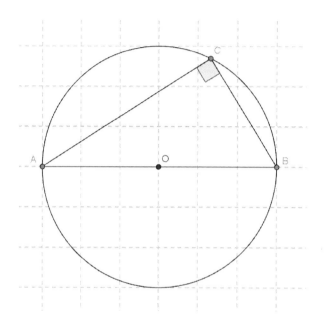

Draw a diameter and angle to circumference.

The angle to the centre from 2 points is now 180 ° hence the angle to the

circumference is half this = 90°, a right angle.

The angle from a diameter to the circumference is a right angle

Tangent Properties

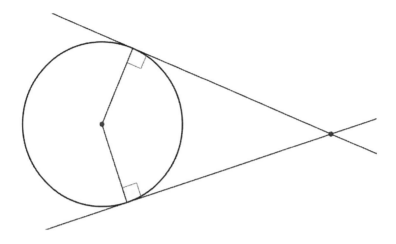

Tangents from a point to a circle touch at one point only

From symmetry, they make a right angle with a radius to that point

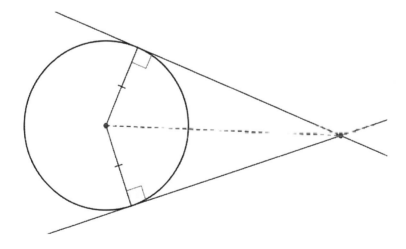

Line drawn as shown makes 2 congruent right angle triangles, equal radii, equal shared hypotenuse (dotted)

Hence tangents from same point to circle are equal

Alternate Segment

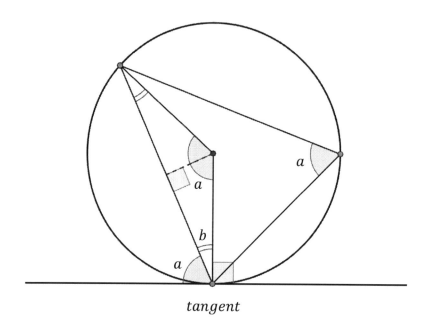

tangent

The angle **a** made with the tangent to a circle and chord of inscribed triangle adds to 90° with angle **b** (tangent properties). The isosceles triangle formed by this chord and the radii to the centre is perpendicularly bisected. The angle at centre bisected also adds to 90° with angle **b** (properties of right angled triangles), so equals angle **a**. This angle, being half of centre angle equals the alternate angle at circumference, also angle **a** (double angle at centre)

Therefore

> The angle made with chord of a circle and tangent equals the angle in the alternate segment

TRIG

Sine Rule

Cosine Rule

Area Triangle using Sine

Sine Rule involving Diameter

Cosine Rule

Sine Rule

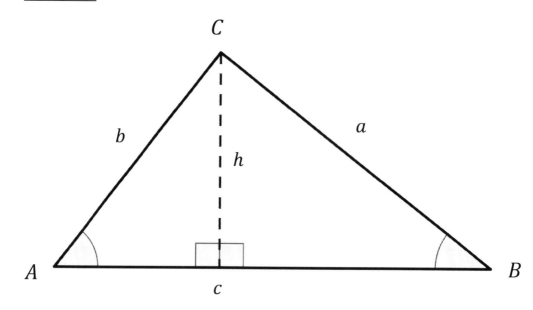

$$Sin\ A = \frac{h}{b} \qquad Sin\ B = \frac{h}{a}$$

$$h = b\ Sin\ A = a\ Sin\ B$$

$$\therefore \quad \frac{Sin\ A}{a} = \frac{Sin\ B}{b} \quad or \quad \frac{a}{Sin\ A} = \frac{b}{Sin\ B}$$

Similarly a perpendicular line can be dropped from A to line CB and the rule proved for angle C also

$$\therefore \quad \frac{a}{Sin\ A} = \frac{b}{Sin\ B} = \frac{c}{Sin\ C}$$

Cosine Rule

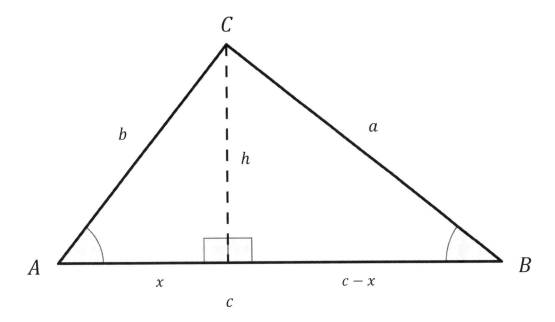

From Pythagoras – left triangle

$$b^2 = x^2 + h^2$$

right triangle

$$a^2 = (c - x)^2 + h^2$$

$$= c^2 - 2xc + x^2 + h^2$$

Substitute $b^2 = x^2 + h^2$ in the above

$$a^2 = b^2 + c^2 - 2xc$$

From left triangle $\quad x = b \, Cos \, A$

$$\therefore \quad a^2 = b^2 + c^2 - 2bc \, Cos \, A$$

Area Triangle using Sine

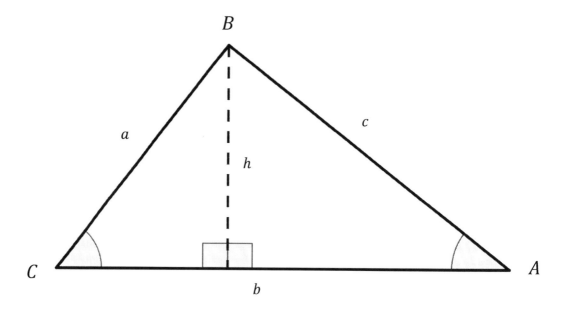

$$Sin\ C = \frac{h}{a} \qquad \therefore \quad h = a \times Sin\ C$$

$$Area\ triangle = \frac{1}{2}\ base\ \times height\ = \frac{1}{2}\ b\ h$$

$$\therefore\ \textbf{\textit{Area triangle}} = \frac{1}{2}\ \textbf{\textit{ab SinC}}$$

Sine Rule involving Diameter

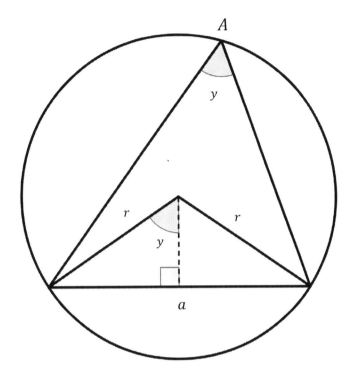

From small right angle triangle made by dropping a perpendicular bisector
from centre to chord length a

$$Sin(y) = \frac{a/2}{r} \qquad \therefore \quad \frac{a}{2} = r\,Sin(y)$$

$$\therefore \quad a = 2r\,Sin(y)$$

Angle A = y

$$\therefore \quad 2r = \frac{a}{Sin\,A}$$

And from Sine Rule

$$\boldsymbol{Diameter} = \boldsymbol{2r} = \frac{\boldsymbol{a}}{\boldsymbol{Sin\,A}} = \frac{\boldsymbol{b}}{\boldsymbol{Sin\,B}} = \frac{\boldsymbol{c}}{\boldsymbol{Sin\,C}}$$

Cosine Rule

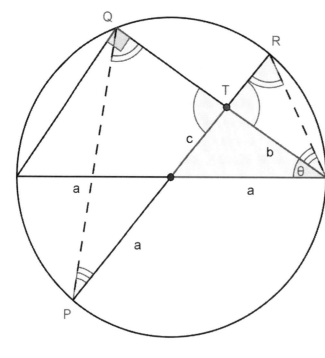

$QT = 2aCos\theta - b$

(right angle triangle from diameter)

$TR = a - c$

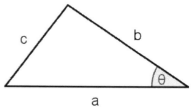

Similar triangles PQT & SRT

(Circle theorem rules and opposite angles)

$$\therefore \frac{QT}{TR} = \frac{PT}{TS}$$

$$\therefore \frac{2a\,Cos\,\theta - b}{a - c} = \frac{a + c}{b}$$

$$\therefore \ (2a\,Cos\theta - b)b \ = \ (a + c)(a - c)$$

$$\therefore \ 2ab\,Cos\theta - b^2 \ = \ a^2 - \ c^2$$

$$\therefore \ \boldsymbol{c^2 \ = \ a^2 + \ b^2 \ - \ 2ab\,Cos\theta}$$

where θ is angle opposite c

PYTHAGORAS

Thabbit ibn Kurrah   10th Century

Bhaskara   12th Century

Perigal   1870's

Liu Hui   250 AD

E.D. Dekker   1888

Classic Pythagoras   6th Century BC

J. Versluys   1914

J.E. Bottcher   1900

Pythagoras   Thabbit ibn Kurrah

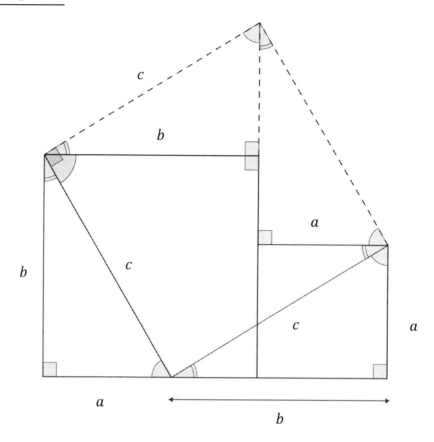

Take two adjoining squares of side a and b , construct two congruent triangles as shown, of sides a , b and hypotenuse c, to point on base

Dotted triangles show new position (can be rotated)

From properties of right angled triangles ◁ plus △ equal 90°

A square of side c is formed

Area

$$a^2 + b^2 = c^2$$

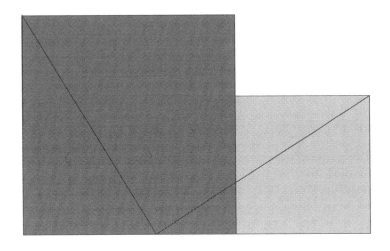

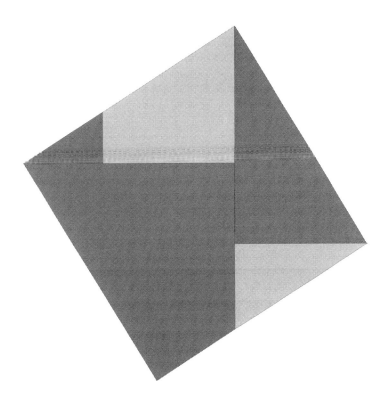

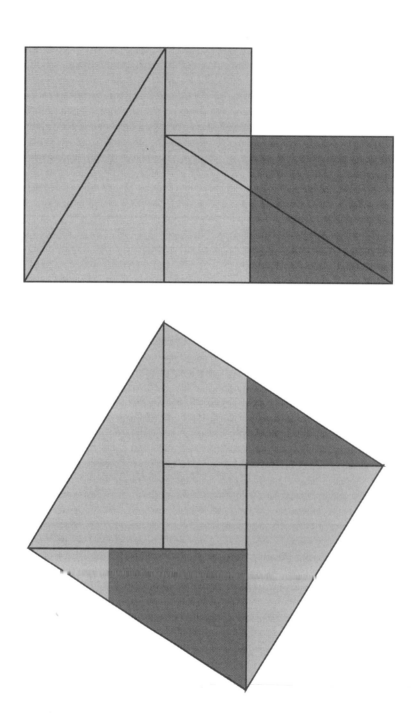

Pythagoras Bhaskara

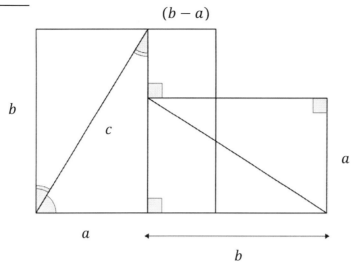

2 squares of side a and b divided into 4 right angled triangles , of sides a , b and hypotenuse c , and square side $(b - a)$ as shown

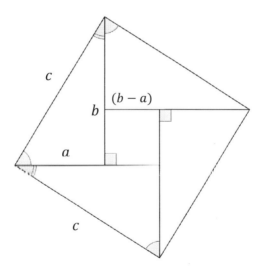

A square of side c is formed – plus equal 90°

Area $\qquad c^2 = (b - a)^2 + 4 \times \frac{1}{2} ab$ .... small square plus 4 triangles

$$c^2 = a^2 + b^2$$

Pythagoras   Perigal

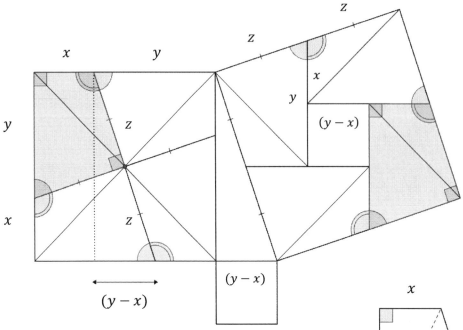

Take the medium square from the 3 drawn from the
sides of a right angled triangle – find its centre (from diagonals)
and draw 2 lines parallel to the large square through this,
forming 4 congruent quadrilaterals of sides x,y and z as shown,

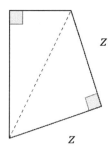

Drop a perpendicular from the point x from top left corner, as shown, forming a congruent
triangle to the original, with base $(y - x)$

The 4 quadrilaterals fit around the small square to form the large square of side 2z –

properties of squares; 4 right angles and equal sides, the angles ◺ plus ◹ equal 180°

Area of congruent quadrilateral $= \dfrac{1}{2}xy + \dfrac{1}{2}z^2$

Area of medium square $(x + y)^2 = 4 \times \dfrac{1}{2}(xy + z^2) = 2xy + 2z^2$ $\therefore$ $\boldsymbol{x^2 + y^2 = 2z^2}$

Sum of 2 smaller squares $(x + y)^2 + (y - x)^2 = 2(x^2 + y^2) = 4z^2$

∴ **the square on the hyponenuse**
= the sum of the squares on the other two sides

68

Perigal   1870's

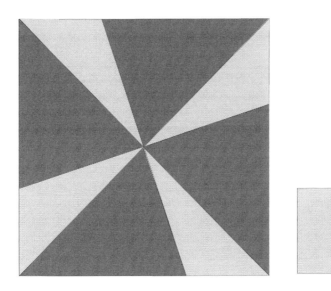

Pythagoras   Liu Hui

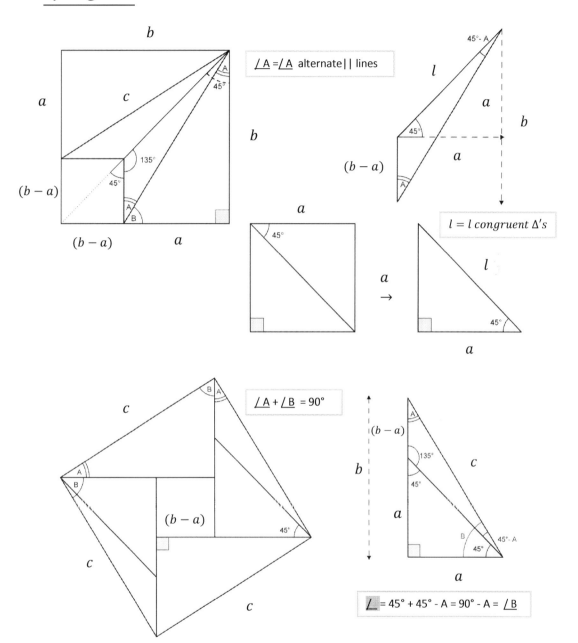

∠A =∠A alternate || lines

∠A + ∠B = 90°

l = l congruent Δ's

∠ = 45° + 45° - A = 90° - A = ∠B

Area of square side c = area of the small square plus 4 triangles

$$c^2 = (b-a)^2 + 4 \times \frac{1}{2}ab = b^2 + a^2 - 2ab + 2ab = a^2 + b^2$$

$$\boxed{c^2 = a^2 + b^2}$$

Pythagoras Symmetry   E.D.Dekker

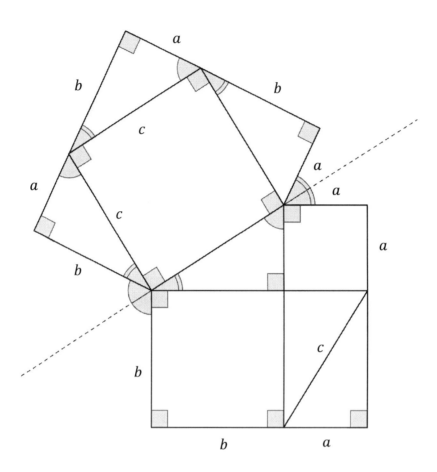

The dotted line of symmetry shows that the large square (of side c) plus 3 congruent right angled triangles (sides a, b and c) mirrors the area of the two smaller squares (sides a and b) plus the same 3 triangles

Therefore when you subtract the 3 triangles from both areas

$$c^2 = a^2 + b^2$$

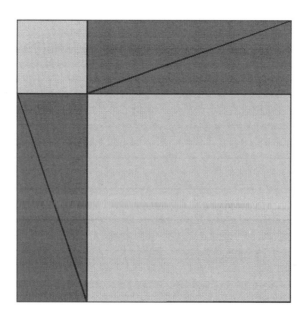

Pythagoras Classic

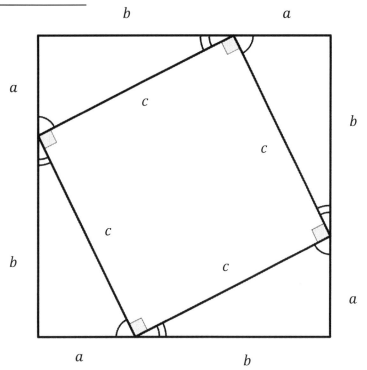

A square of side a + b is drawn, with square side c constructed within as shown. Four congruent right angled triangles are formed with sides a,b and c

Properties right angled triangle : angles ◿ ◹ sum to 90°

Square c^2 has 4 angles 90° and four equal sides

Area of large square = area of 4 triangles plus smaller square

$$(a + b)^2 = 4 \times \frac{1}{2}ab + c^2$$
$$a^2 + b^2 + 2ab = 2ab + c^2$$

Subtract 2ab from both sides

$$\therefore \quad \boldsymbol{a^2 + b^2 = c^2}$$

Pythagoras   J.Versluys

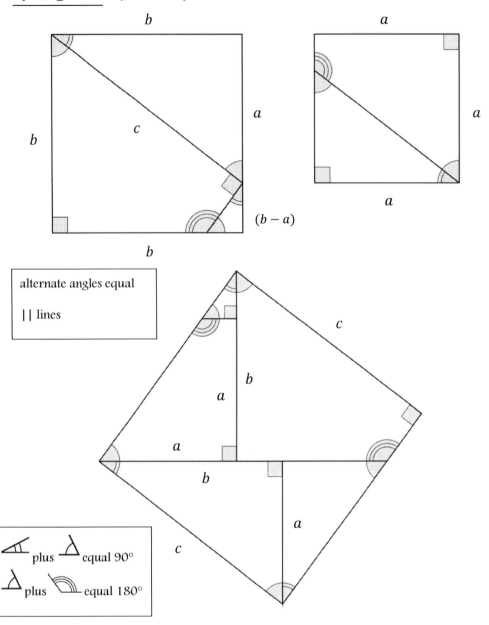

alternate angles equal

|| lines

plus equal 90°

plus equal 180°

Two squares of side b and a are dissected as shown forming 3 similar triangles, one of sides a, b and c, and 2 similar quadrilaterals . These are rearranged as shown to form a square side c

$$a^2 + b^2 = c^2$$

J. Versluys   1914

Pythagoras   J.E.Bottcher

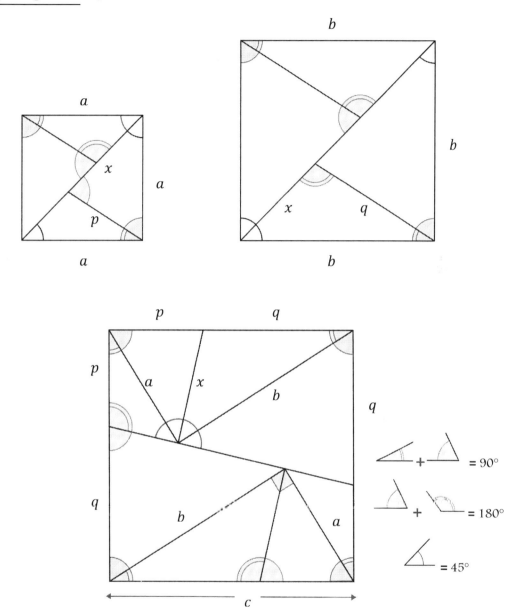

2 squares of side a and b are dissected as shown, a diagonal creates 45° angles.

They are split into 2 pairs of similar triangles , (x marked on diagonal both squares),

and rearranged to make square side $p + q$.

A right angled triangle of side a , b and (p + q) is formed. Let p + q = c

$$a^2 + b^2 = c^2$$

EXTRAS

Pentagram Star

Quadratic Roots

Power Laws

Density / Speed

Vectors – the One Rule

Pythagoras from Similar Triangles or Tan θ

Pentagram – Golden Ratio

Pentagram Star

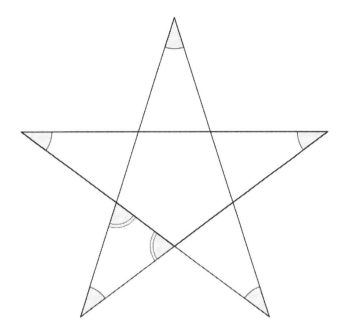

From exterior angle of a triangle equals the sum of the two opposite angles –
each double stripe angle equals two star point angles – Therefore one triangle
encloses 5 of the star point angles

Sum of angles in a 5 pointed star equals 180°

Quadratic Roots

$$ax^2 + bx + c = 0$$

$\times$ *both sides by* $4a$

$$4a^2x^2 + 4abx + 4ac = 0 \qquad -eq(i)$$

and

$$(2ax + b)^2 = 4a^2x^2 + 4abx + b^2$$

$$\therefore 4a^2x^2 + 4abx = (2ax + b)^2 - b^2$$

Substitute RHS in above $-eq\,(i)$

$$(2ax + b)^2 - b^2 + 4ac = 0$$

$$\therefore (2ax + b)^2 = b^2 - 4ac$$

Take square root of both sides

$$2ax + b = \pm\sqrt{b^2 - 4ac}$$

$$\therefore 2ax = -b \pm \sqrt{b^2 - 4ac}$$

$$\therefore x = \frac{-b + \sqrt{b^2 - 4ac}}{2a}$$

Power Laws

$$x^3 \times x^2 = x^{3+2} = x^5 \qquad x \times x \times x \quad \times x \times x = x^5 \qquad \text{add powers}$$

$$x^7 \div x^4 = x^{7-4} = x^3 \qquad \frac{x \times x \times x \times x \times x \times \cancel{x} \times \cancel{x} \times \cancel{x} \times \cancel{x}}{\cancel{x} \times \cancel{x} \times \cancel{x} \times \cancel{x}} = x^3 \text{ subtract powers}$$

$$(x^2)^3 = x^{2\times3} = x^6 \qquad (x^2 \times x^2 \times x^2 = x^6) \qquad \text{multiply powers}$$

$$x^0 = 1 \qquad \frac{x}{x}(=1) = x^1 \div x^1 = x^{1-1} = x^0 \qquad \text{any number to power}^0 = 1$$

$$x^{-n} = \frac{1}{x^n} \qquad x^0 \div x^3 = x^{0-3} = x^{-3} = \frac{1}{x^3} \text{ negative powers are reciprocal}$$

$$x^{\frac{1}{2}} = \sqrt{x} \qquad x^{\frac{1}{2}} \times x^{\frac{1}{2}} = x \qquad \sqrt{x} \times \sqrt{x} = x \qquad \text{fractional powers are roots}$$

Fractional and Negative Indices

$$16^{-\frac{3}{4}} = \frac{1}{16^{\frac{3}{4}}} = \frac{1}{\left(16^{\frac{1}{4}}\right)^3} = \frac{1}{2^3} = \frac{1}{8}$$

$$\left(\frac{8}{27}\right)^{-\frac{2}{3}} = \left(\frac{27}{8}\right)^{\frac{2}{3}} = \left(\frac{27^{\frac{1}{3}}}{8^{\frac{1}{3}}}\right)^2 = \left(\frac{3}{2}\right)^2 = \frac{9}{4}$$

Surds

$$\sqrt{48} = \sqrt{16 \times 3} = \sqrt{16} \times \sqrt{3} = 4\sqrt{3}$$

$$\sqrt{\frac{48}{75}} = \frac{\sqrt{48}}{\sqrt{75}} = \frac{4\sqrt{3}}{\sqrt{25 \times 3}} = \frac{4\sqrt{3}}{\sqrt{25}\sqrt{3}} = \frac{4}{5}$$

To rationalise (not dividing by irrational) $\dfrac{6}{\sqrt{3}} = \dfrac{6}{\sqrt{3}} \times \dfrac{\sqrt{3}}{\sqrt{3}} = \dfrac{6\sqrt{3}}{3} = 2\sqrt{3}$

And $\dfrac{1}{2+\sqrt{3}} = \dfrac{1}{(2+\sqrt{3})} \times \dfrac{(2-\sqrt{3})}{(2-\sqrt{3})} = \dfrac{2-\sqrt{3}}{2-3} = \dfrac{2-\sqrt{3}}{-1} = \sqrt{3}-2$

Density

Density is the measurement of the amount of matter in a volume

$$\therefore \ Density \ = \ \frac{Mass}{Volume}$$

Common Densities :=   *Examples given in* $10^3 kg/m^3$ *or* g/cm^3

Water	– 1.0	Ice	– 0.917
Copper	– 8.95	Air(sea level)	– 0.001225
Helium	– 0.0001785 (1.785×10^{-4})		
Glass	– 2.6	Aluminium	– 2.70
Coal	– 1.35	Oak	– 0.72
Gold	– 19.29	ρ (rho) is the symbol for density	

The concept of density was responsible for Archimedes 'eureka' moment.
He realised that the volume of an object is given by its displacement in water
(in his bath!) so he could compare the densities and purity of precious metals
Units also give the formula - $eg. \ kg/m^3 \ = \$ mass per volume

Speed

Speed is the distance travelled over time – as seen by units eg. km/hr

$$Speed \ = \ \frac{Distance}{Time}$$

Vectors – the One Rule

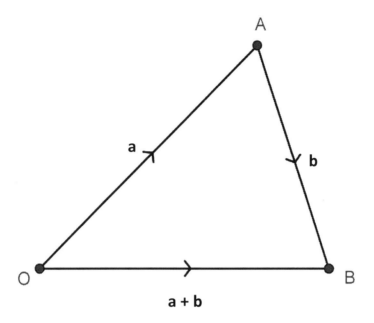

A vector is a direction – if you start at one point and end at another, the paths (vectors) you take on route are equal

If you start at O and travel to B , via A, through vectors a and b this is equal to the short route $\overrightarrow{OB}$

$$\overrightarrow{OA} + \overrightarrow{AB} = \overrightarrow{OB}$$

$$\overrightarrow{OB} = a + b$$

Pythagoras from Similar Triangles or Tan θ

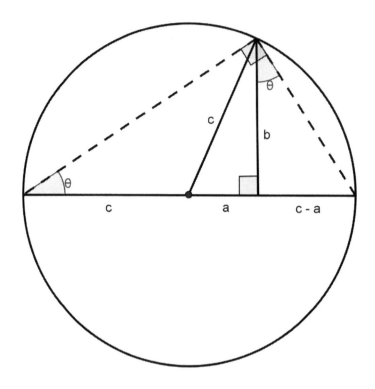

From similar triangles – or $\tan \theta$

$$\frac{b}{c+a} = \frac{c-a}{b}$$

$$b^2 = (c+a)(c-a)$$

$$b^2 = c^2 - a^2$$

$$\therefore \quad c^2 = a^2 + b^2$$

Pentagram – Golden Ratio

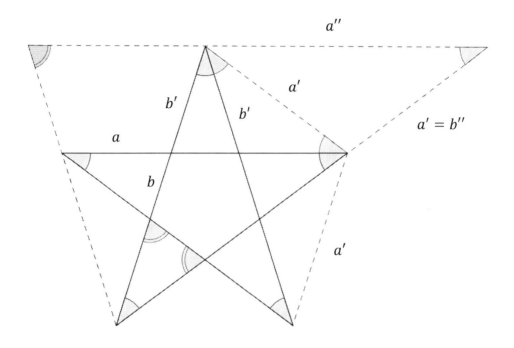

In a regular pentagram

The ratio $\dfrac{a}{b} = \dfrac{a'}{b'}$ and $\dfrac{a'}{b'} = \dfrac{a''}{b''}$ from similar triangles

$a' = a + b$ and $a'' = a' + b'$

$b' = a$ and $b'' \quad a'$

$\therefore \dfrac{a}{b} = \dfrac{a+b}{a}$ which is the Golden Ratio

as are $\dfrac{a'}{b'}$ and $\dfrac{a''}{b''}$

$$\boldsymbol{\dfrac{a}{b} = \varphi \; Golden \; ratio}$$

$$\varphi = \frac{a}{b} = \frac{a+b}{a} = 1 + \frac{b}{a}$$

$$\therefore \varphi = 1 + \frac{1}{\varphi}$$

$$\therefore \varphi^2 - \varphi - 1 = 0$$

$$\varphi = \frac{1 \pm \sqrt{5}}{2}$$

"I THINK YOU SHOULD BE MORE EXPLICIT HERE IN STEP TWO."

RADIANS & CIRCLES

Radian Definition

Length Arc Circle, Area Sector in Radians

Area of Segment of Circle in Radians

Radian Definition

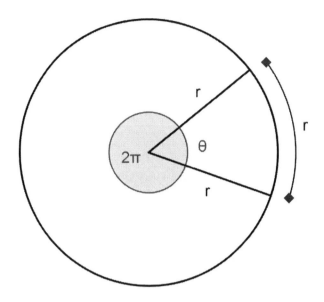

A radian is the angle subtended at the centre of a circle by an arc of length r (radius of circle)

$$\theta = 1^c \quad (\, one\ radian\,)$$

If an angle is given without degree symbol take it to be radian

And in degrees :

$$1\ radian = \frac{360°}{2\pi}$$

$$= \frac{180°}{\pi} = 57.29577951\dots°$$

And

$$1° = \frac{\pi}{180}\ radians$$

Length Arc Circle, Area Sector in Radians

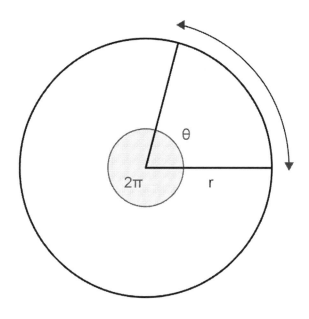

$$length\ arc\ =\ \frac{\theta}{2\pi}\ \times\ 2\pi r$$

$$=\ \boldsymbol{r\theta}$$

$$area\ sector\ =\ \frac{\theta}{2\pi}\ \times\ \pi r^2$$

$$=\ \boldsymbol{\frac{1}{2}r^2\theta}$$

Area of Segment of Circle in Radians

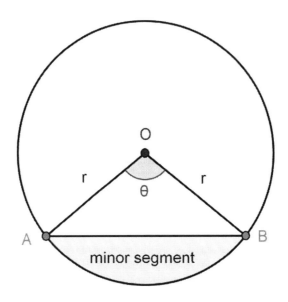

Area minor segment = area sector AOB – area triangle AOB

$$Area\ sector\ AOB\ =\ \frac{1}{2}\,r^2\theta$$

$$Area\ triangle\ AOB\ =\ \frac{1}{2}\,r^2\,sin\theta$$

$$\therefore\ \boldsymbol{Area\ segment} = \frac{1}{2}r^2(\,\theta - sin\theta\,)$$

LOGARITHMS

Logarithms

Log Rules OK

Change of Base Log Rule

Logarithms

What is a log ?

$$log_{base} \ number \ = \ power$$

$$log_{10} 100 \ = \ 2$$

$$10^2 \ = \ 100$$

$$log_{base} \ number \ = \ power$$

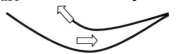

$$base^{power} \ = \ number$$

$$b^p \ = \ n$$

$$log_b n = p$$

$$log_a a \ = \ 1$$

$$log_a 1 \ = \ 0$$

$$log_a a^2 = 2 \quad log_a a^3 = 3 \quad log_a a^4 = 4 \quad \text{(logs can substitute for numbers)}$$

$$log_e e = 1 \quad (\ln e = 1) \qquad log_e 1 = 0 \quad (\ln 1 = 0) \quad \ln = \text{natural logarithm}$$

Log Rules OK

$$\log_b x = p \qquad \text{and} \qquad \log_b y = q$$

$$b^p = x \qquad \text{and} \qquad b^q = y$$

$$xy = b^p \times b^q \qquad \therefore xy = b^{p+q}$$

$$\therefore \log_b xy = p + q$$

$$\boxed{\therefore \ \boldsymbol{\log_b xy = \log_b x + \log_b y}}$$

Likewise

$$\frac{x}{y} = b^{p-q}$$

$$\therefore \log_b \frac{x}{y} = p - q$$

$$\boxed{\therefore \ \boldsymbol{\log_b \frac{x}{y} = \log_b x - \log_b y}}$$

And

$$x^k = (b^p)^k = b^{pk}$$

$$\therefore \log_b x^k = pk$$

$$\boxed{\therefore \ \boldsymbol{\log_b x^k = k \log_b x}}$$

Change of Base Log Rule

$$log_b x = p$$

$$b^p = x$$

$$log_m b^p = log_m x$$

$$p \, log_m b = log_m x$$

$$\therefore \; p = \frac{log_m x}{log_m b}$$

$$\therefore \; \boldsymbol{log_b x = \frac{log_m x}{log_m b}}$$

SERIES & SUMS

Sum of Geometric Series

Sum of Geometric Series to Infinity

Sum of Geometric Series(algebra)

Sum of Integers – Triangular Numbers

Sum of Arithmetic Progression

Sum of Squares

Sum of Odd Integers

Consecutive Odd Numbers Make Cubes

Sum of Cubes

Sum of Geometric Series

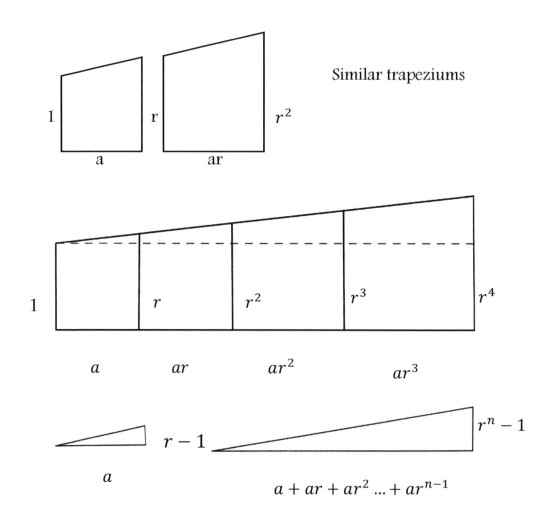

Similar trapeziums

Similar triangles

$$\frac{a + ar + ar^2 \ldots + ar^{n-1}}{a} = \frac{r^n - 1}{r - 1}$$

$$S_n = \frac{a(r^n - 1)}{r - 1} = \frac{a(1 - r^n)}{1 - r}$$

Sum of Geometric Series to Infinity

$|r| < 1$

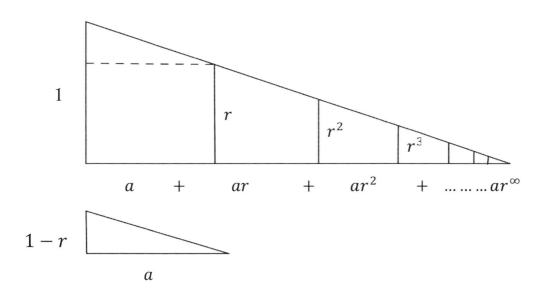

Similar triangles

$$\frac{a + ar + ar^2 \ldots + ar^\infty}{1} = \frac{a}{1 - r}$$

$$S_\infty = \frac{a}{1 - r}$$

Sum of Geometric Series(algebra)

$$S_n = a + ar + ar^2 + ar^3 + \ldots ar^{n-1}$$

$$rS_n = ar + ar^2 + ar^3 + \ldots ar^{n-1} + ar^n$$

$$S_n - rS_n = a - ar^n = a(1 - r^n)$$

$$S_n(1 - r) = a(1 - r^n)$$

$$\boxed{S_n = \frac{a(1 - r^n)}{1 - r}}$$

$$\text{If } |r| < 1 \qquad r^n \to 0$$

$$\boxed{S_\infty = \frac{a}{1 - r}}$$

Sum of Integers – Triangular Numbers

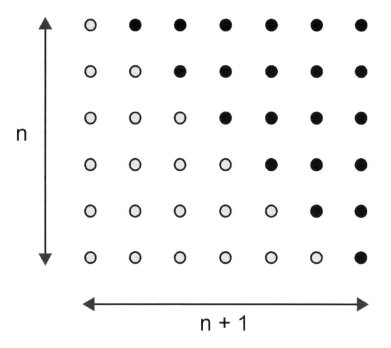

Sum of integers $1 + 2 + 3 + 4 + \ldots \ldots n$

$$= \frac{n}{2}(n+1)$$

Sum of Arithmetic Progression

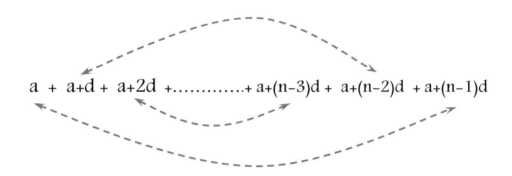

$$a \ + \ a{+}d \ + \ a{+}2d \ +\ldots\ldots\ldots+ a{+}(n{-}3)d \ + \ a{+}(n{-}2)d \ + a{+}(n{-}1)d$$

The terms add in pairs to make $\dfrac{n}{2}$ *terms of* $2a + (n-1)d$

$$\therefore \ \textit{Sum of arithmetic progression} = \dfrac{\boldsymbol{n}}{\boldsymbol{2}} \left[\, 2\boldsymbol{a} + (\boldsymbol{n} - 1)\boldsymbol{d} \,\right]$$

$$or \ \dfrac{\boldsymbol{n}}{\boldsymbol{2}} \, (\, \boldsymbol{a} + \boldsymbol{l} \,) \qquad l = \textit{last term}$$

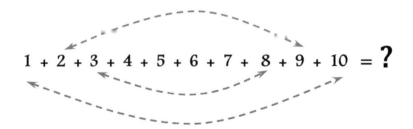

$$1 \ + \ 2 \ + \ 3 \ + \ 4 \ + \ 5 \ + \ 6 \ + \ 7 \ + \ 8 \ + \ 9 \ + \ 10 \ = \ \mathbf{?}$$

Sum of Squares

$$2n + 1$$

$$\frac{n}{2}(n+1)$$

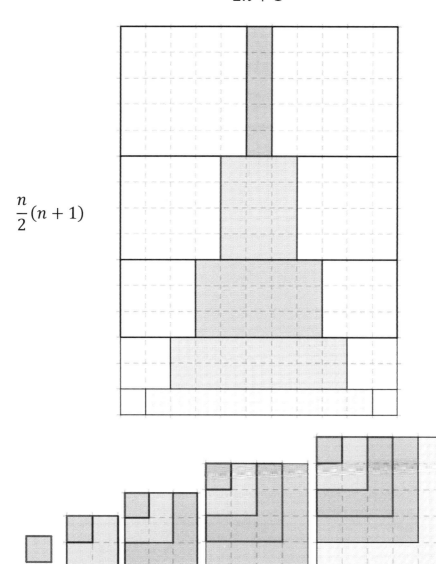

$$3 \; sets \; of \; square \; sums = \frac{n}{2}(n+1)(2n+1)$$

$$\therefore \; \textbf{\textit{Sum of Squares}} = \frac{n}{6}(n+1)(2n+1)$$

Sum of Odd Integers

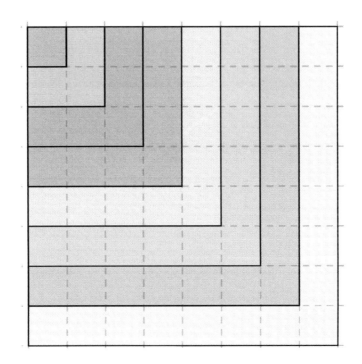

Sum of n odd integers $= n^2$

Consecutive Odd Numbers Make Cubes

1

3+5

7+9+11

13+15+17+19

21+23+25+27+29

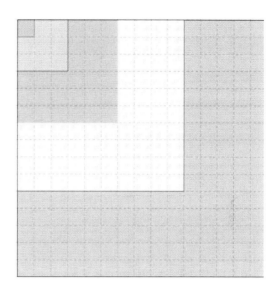

The mid point of the sequences of these continuous consecutive n odd numbers is n^2

The other numbers average to make a n lots of $n^2 = n^3$

Therefore the sum of a continuous sequence of n consecutive numbers $= n^3$

Sum of Cubes

$$\frac{n}{2}(n+1)$$

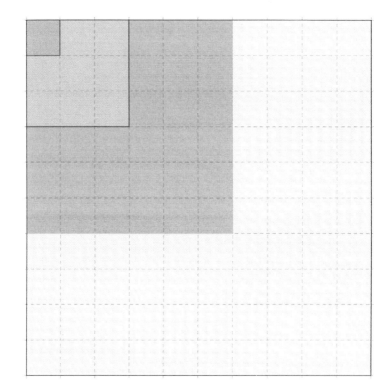

$$\frac{n}{2}(n+1)$$

$$\text{Sum of cubes} = \left(\frac{n}{2}(n+1)\right)^2$$

CALCULUS

Differentiation of a Function

Graph and Derivative of e^x

Cos x and its Derivative $= -$ Sin x

Differentiation Chain Rule Using u

Differentiation Product Rule

Differentiation Quotient Rule

Differentiation of ln(x)

Differentiation Power of x

Integration by Parts

Trapezium Rule

Integration as Area and Volume

Differentiation of a Function

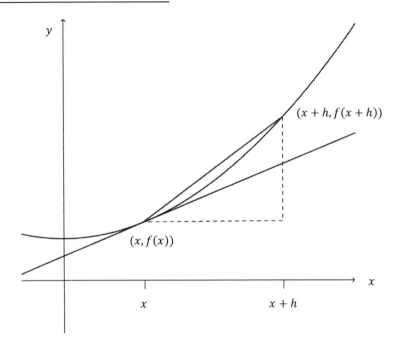

A gradient of a function is the rate of change of the function with respect to x.

We find this by finding the gradient of the chord near the tangent , and

reducing this in length till it is in effect the same gradient as the tangent

$$gradient = \frac{diff\ in\ y}{diff\ in\ x} = \frac{f(x+h) - f(x)}{h}$$

$$So\ for\ \ y = x^2 \quad gradient = \frac{(x+h)^2 - x^2}{h}$$

$$= \frac{x^2 + 2xh + h^2 - x^2}{h} \quad = \quad 2x + h$$

As $h \to 0$ $\quad$ *gradient of tangent to* $x^2 = 2x$

$$\boxed{\quad Rule\ y = x^n \qquad \frac{dy}{dx} = nx^{n-1} \quad}$$

The reverse process is to integrate $\quad \int \frac{dy}{dx} dx = y + c$

$$or\ \ if\ y = x^n \quad \int y\ dx = \frac{x^{n+1}}{n+1} + c$$

Graph and Derivative of e^x

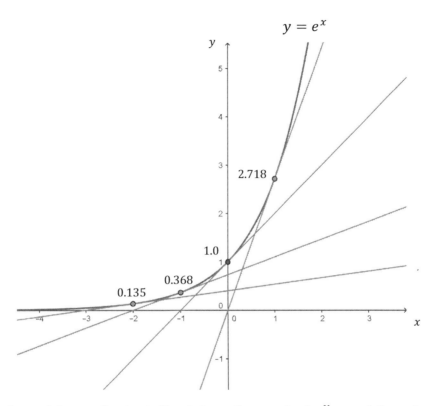

The values of the gradients at all points on the graph of e^x equal the values of the function at that point $\therefore f'(x) = f(x)$ for e^x

The special natural growth constant e can be defined from continuous compound interest

$$e = \left(1 + \frac{1}{n}\right)^n \; as\; n \to \infty$$

The value of e becomes constant at 2.718 at $n = 10^4$

e is used as the base for natural logarithms $\log_e = \ln$

$$y = e^x \quad \frac{dy}{dx} = e^x \; and \; \int e^x \, dx = e^x + c$$

Cos x and its Derivative $= -$ Sin x

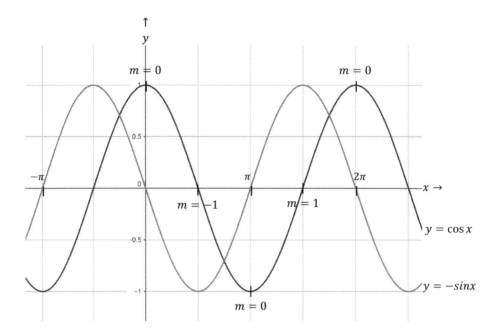

The values of the gradient of $\cos x$ equal the values of $-\sin x$

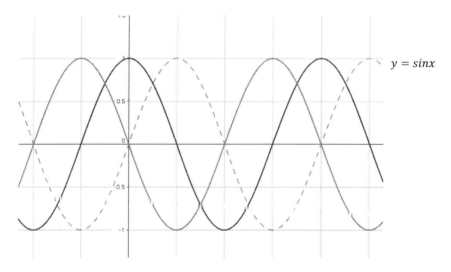

It can be seen that $-\sin x$ is the inversion of $\sin x$ (the dotted line) in the x axis,   and that the derivative of $\sin x$ will be the graph of $\cos x$

Differentiation Chain Rule Using u

With composite functions $\qquad y = f(g(x))$

$$y = f(u) \qquad substituting \; u \; for \; g(x),$$

$$\frac{dy}{du} \times \frac{du}{dx} = \frac{dy}{dx}$$

$$\boxed{\frac{dy}{dx} = \frac{dy}{du} \times \frac{du}{dx}}$$

example

$$y = sin^2 x \qquad let \; \mathbf{u} = sin \, x \qquad \therefore \;\; y = u^2$$

$$\frac{du}{dx} = \cos x \qquad \frac{dy}{du} = 2u$$

$$\therefore \quad \frac{dy}{dx} = \frac{dy}{du} \times \frac{du}{dx} = 2u \cos x = 2 \sin x \cos x$$

from chain rule these rules are obtained

$$y = e^{kx} \quad \frac{dy}{dx} = ke^{kx} \, ; \qquad y = \sin kx \quad \frac{dy}{dx} = k \cos kx$$

and integration by reverse differentiation

$$\int e^{kx} dx = \frac{e^{kx}}{k} + c \quad and \quad \int \cos kx \, dx = \frac{\sin kx}{k} + c$$

Differentiation Product Rule

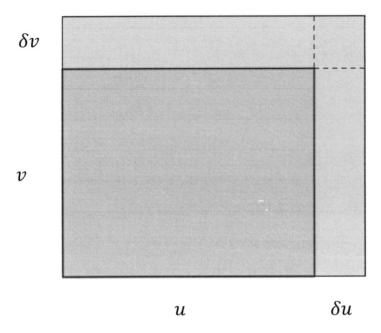

Take rectangle sides u,v and area y, small increment change δu & δv

As δu & $\delta v \rightarrow 0$ the difference in area rectangle uv and resulting rectangle will be

$$\delta y = u\,\delta v + v\,\delta u \;\; since \;\; \delta u \delta v \rightarrow 0$$

$$\therefore \;\; \frac{\delta y}{\delta x} = u\frac{\delta v}{\delta x} + v\frac{\delta u}{\delta x}$$

$$As\; \delta x \rightarrow 0 \;\; \frac{\delta y}{\delta x} \rightarrow \frac{dy}{dx}, \;\;\; \frac{\delta v}{\delta x} \rightarrow \frac{dv}{dx}, \;\;\; \frac{\delta u}{\delta x} \rightarrow \frac{du}{dx}$$

$$\frac{d}{dx}uv = u\frac{dv}{dx} + v\frac{du}{dx}$$

Differentiation Quotient Rule

$$y = \frac{u}{v} \quad where\ u\ \&\ v\ are\ functions\ of\ x$$

$$Use\ product\ rule \quad substitute \ \ \frac{1}{v} \ \ for \ \ v$$

$$\frac{d}{dx}\left(\frac{u}{v}\right) \ = \ u\frac{d}{dx}\left(\frac{1}{v}\right) + \left(\frac{1}{v}\right)\frac{du}{dx}$$

$$= \ -\frac{u}{v^2}\frac{dv}{dx} + \frac{1}{v}\frac{du}{dx}$$

$$\therefore \ \frac{d}{dx}\left(\frac{u}{v}\right) = \frac{v\dfrac{du}{dx} - u\dfrac{dv}{dx}}{v^2}$$

Differentiation of ln(x)

$$y = \ln x$$

$$x = e^y$$

$$\frac{dx}{dy} = e^y$$

$$\frac{dy}{dx} = \frac{1}{\frac{dx}{dy}}$$

$$\frac{dy}{dx} = \frac{1}{e^y}$$

$$\therefore \frac{dy}{dx} = \frac{1}{x}$$

Integration by reverse differentiation

$$\int \frac{1}{x} dx = y + C = \ln(x) + C$$

Differentiation Power of x

$$y = a^x$$

Take natural logs both sides

$$\ln y = \ln a^x$$

$$\ln y = x \ln a$$

Differentiate both sides with respect to x, a is a constant,

chain rule for ln(y)

$$\frac{1}{y}\frac{dy}{dx} = \ln a$$

$$\frac{dy}{dx} = y \ln a$$

$$\therefore \frac{dy}{dx} = a^x \ln a$$

Integration by reverse differentiation

$$\int a^x \, dx = \frac{y}{\ln a} + C = \frac{a^x}{\ln a} + C$$

Integration by Parts

From the product rule

$$\frac{d}{dx}uv = u\frac{dv}{dx} + v\frac{du}{dx}$$

Integrate all with respect to x

$$\int \frac{d}{dx}uv\ dx = \int u\frac{dv}{dx}\ dx + \int v\frac{du}{dx}\ dx$$

LHS reverse differentiation

$$uv = \int u\frac{dv}{dx}\ dx + \int v\frac{du}{dx}\ dx$$

Rearrange

$$\int u\frac{dv}{dx}\ dx = uv - \int v\frac{du}{dx}\ dx$$

Trapezium Rule

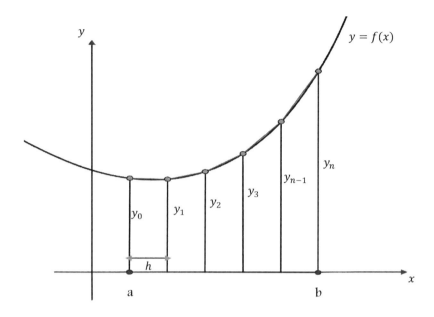

To find area under curve by approximation, when integration is not possible. Equally spaced (h) trapeziums are drawn from curve to x axis. Each trapezium has area $\frac{1}{2}h\,(y_i + y_k)$

$$\int_a^b y\,dx = \frac{1}{2}h\,[\,y_0 + 2(y_1 + y_2 \ldots + y_{n-1}) + y_n\,]$$

$$h = \frac{b-a}{n} \qquad and \quad y_i = f(a+ih)$$

Integration as Area and Volume

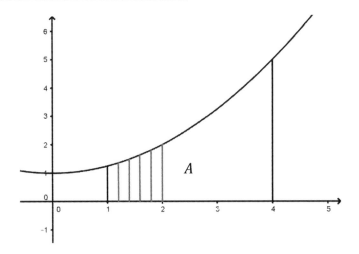

$\int_1^4 y\,dx$ integrating y within limits of x can be thought of as a sum of infinite, infinitely thin 'lines' height y width dx to give an area

The curve can be seen as the rate of change of the area $y = \dfrac{dA}{dx}$ $\therefore A = \int y\,dx$

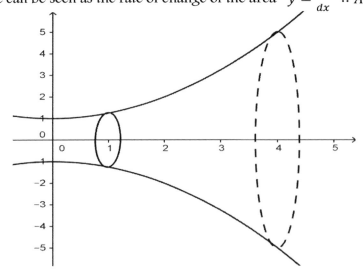

The volume of line y rotated around the x axis can be thought of as a sum of infinite, infinitely thin 'circles' radius y, area πy^2 depth dx

$$\pi \int_1^4 y^2 dx = volume$$

118

VECTORS

Position Vector

The Dot Product

Position Vector

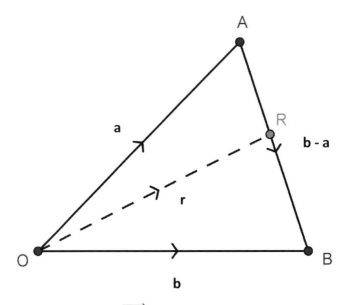

Point R is on vector $\overrightarrow{AB}$,

   t is a scalar which gives its position from A

Position vector $\boldsymbol{r} = \boldsymbol{a} + t(\boldsymbol{b} - \boldsymbol{a})$

Parallel vectors

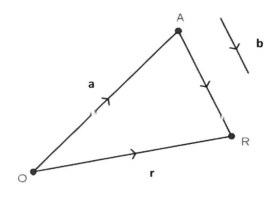

Position vector $\boldsymbol{r} = \boldsymbol{a} + m\boldsymbol{b}$

The Dot Product

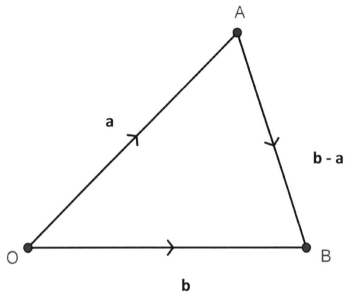

From the Cosine rule

$$Cos\ \theta\ =\ \frac{|\boldsymbol{a}|^2\ +\ |\boldsymbol{b}|^2\ -\ |\boldsymbol{b}-\boldsymbol{a}|^2}{2\ |\boldsymbol{a}|\ |\boldsymbol{b}|}$$

$$|\boldsymbol{a}|\ =\ \sqrt{a_1^2+\ a_2^2+\ a_3^2}$$

$$|\boldsymbol{a}|^2\ +\ |\boldsymbol{b}|^2\ -\ |\boldsymbol{b}-\boldsymbol{a}|^2\ =\ a_1^2+a_2^2+a_3^2\ +\ b_1^2+b_2^2+b_3^2$$
$$-[(b_1-a_1)^2\ +\ (b_2-a_2)^2\ +\ (b_3-a_3)^2]$$

$$=\ a_1^2+a_2^2+a_3^2\ +\ b_1^2+b_2^2+b_3^2$$
$$-\ [a_1^2+a_2^2+a_3^2\ +\ b_1^2+b_2^2+b_3^2-2a_1b_1-2a_2b_2-2a_3b_3\ ]$$

$$\therefore \quad Cos\ \theta = \frac{2(a_1 b_1 + a_2 b_2 + a_3 b_3)}{2\ |a|\ |b|}$$

$$\therefore \quad Cos\ \theta = \frac{a \cdot b}{|a|\ |b|}$$

From this properties of perpendicular and parallel lines are obtained

$$If\ \theta = 90°\ \ Cos\ \theta = 0$$

$$\therefore for\ perpendicular\ lines\ \ a \cdot b\ = 0$$

$$If\ \theta = 0\ \ Cos\ \theta = 1$$

$$\therefore for\ parallel\ lines\ \ a \cdot b = \ |a|\ |b|$$

$$and\ a \cdot a = \ |a|^2$$

Also from this properties of unit vectors (magnitude one) are found

Perpendicular $\quad i \cdot j\ =\ j \cdot k\ =\ i \cdot k\ =\ 0$

Parallel $\quad\quad\quad i \cdot i\ =\ j \cdot j\ =\ k \cdot k\ = 1$

TRIG

Double Angle Formula

Sum of Sines by Area

Sum of Sin & Cos

Angle Difference of Sin, Cos

Double Angle Sine

Double Angle Cosine

Sin Cos Relationship

Exact Values Cos Sin Tan

Adding or Subtracting Different Sines or Cosines

Sums of different Sines and Cosines

Differences of Sines and Cosines

Double Angle Formula

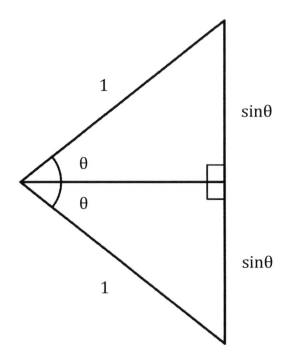

From Cosine Rule

$$Cos\,2\theta \; = \; \frac{1^2 + 1^2 - (2Sin\theta)^2}{2 \times 1 \times 1} \; = \; \frac{2 - 4Sin^2\theta}{2}$$

$$\therefore \quad Cos\,2\theta \; = \; 1 - 2Sin^2\theta$$

Sum of Sines by Area

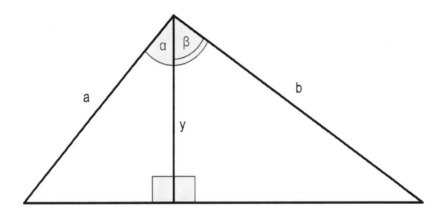

$$y \;=\; a\cos\alpha \;=\; b\cos\beta$$

Area of whole triangle equals sum of the area of 2 smaller triangles

$$\frac{1}{2}ab\sin(\alpha+\beta) \;=\; \frac{1}{2}ay\sin\alpha \;+\; \frac{1}{2}by\sin\beta$$

$$=\frac{1}{2}ab\cos\beta\,\sin\alpha \;+\; \frac{1}{2}ab\cos\alpha\,\sin\beta$$

$\div$ *through by* $\dfrac{1}{2}ab$

$$\boxed{\;\sin(\alpha+\beta) \;=\; \cos\beta\,\sin\alpha \;+\; \cos\alpha\,\sin\beta\;}$$

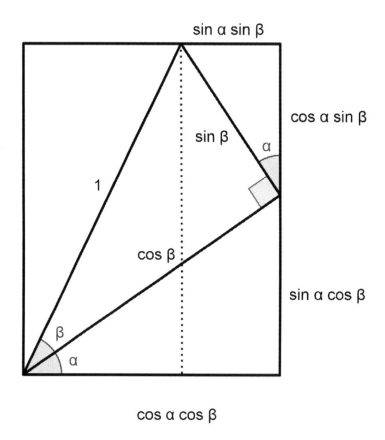

$$Sin\ (\ \alpha + \beta\ ) = \ Cos\ \alpha\ Sin\ \beta\ +\ Sin\ \alpha\ Cos\ \beta$$

$$Cos\ (\ \alpha + \beta\ ) = \ Cos\ \alpha\ Cos\ \beta\ -\ Sin\ \alpha\ Sin\ \beta$$

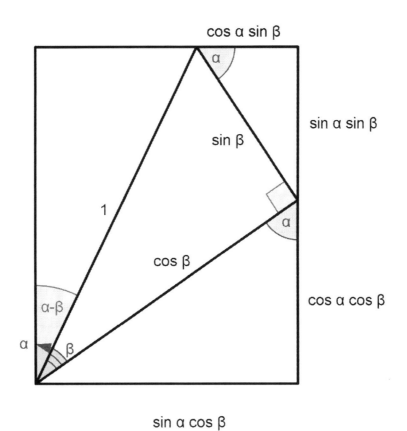

cos α sin β

sin α sin β

sin β

1

cos β

α

α

α-β

α β

cos α cos β

sin α cos β

$$Sin\,(\,\alpha - \beta\,) \;=\; Sin\,\alpha\,Cos\,\beta \;-\; Cos\,\alpha\,Sin\,\beta$$

$$Cos\,(\,\alpha - \beta\,) \;=\; Sin\,\alpha\,Sin\,\beta \;+\; Cos\,\alpha\,Cos\,\beta$$

Double Angle Sine

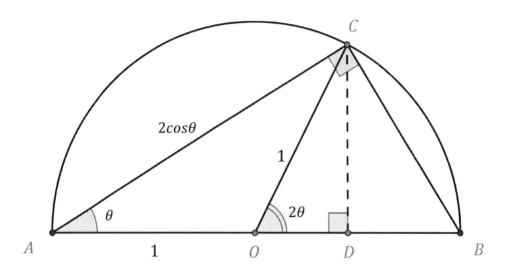

Triangle ABC inscribed in semicircle radius 1

$$AC = 2cos\theta \quad CD = sin2\theta \quad CB = 2sin\theta$$

From similar triangles ACB and ADC or $sin\theta$

$$\frac{CD}{AC} = \frac{BC}{AB}$$

$$\frac{Sin\ 2\theta}{2\cos\theta} = \frac{2\sin\theta}{2}$$

$$Sin\ 2\theta = 2\sin\theta\cos\theta$$

Double Angle Cosine

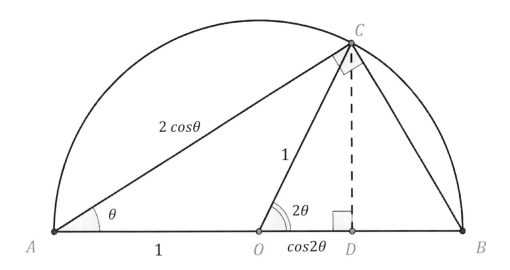

Triangle ABC inscribed in circle radius 1

From similar triangles ACB and ADC or $cos\theta$

$$\frac{AD}{AC} = \frac{AC}{AB}$$

$$\frac{(1 + \cos 2\theta)}{2\cos\theta} = \frac{2\cos\theta}{2}$$

$$Cos\,2\theta = 2\cos^2\theta - 1$$

Sin Cos Relationship

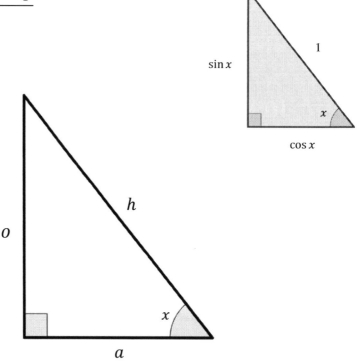

$$\cos x = \frac{a}{h} \qquad\qquad \sin x = \frac{o}{h}$$

$$a = h\ \cos x \qquad\qquad o = h\ \sin x$$

From Pythagoras
$$a^2 + o^2 = h^2$$

$$h^2 \cos^2 x + h^2 \sin^2 x = h^2$$

Divide through by h^2

$$\boxed{\mathbf{\cos^2 x + \sin^2 x = 1}}$$

Exact Values Cos Sin Tan

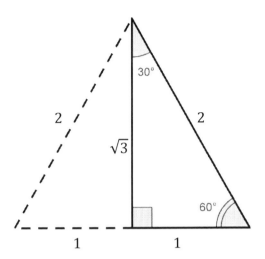
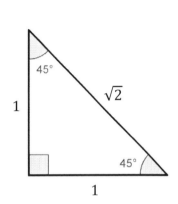

$$Cos\ 60° \ = \ Sin\ 30° \ = \ \frac{1}{2}$$

$$Cos\ 45° \ = \ Sin\ 45° = \frac{1}{\sqrt{2}}$$

$$Sin\ 60° \ = \ Cos\ 30° \ = \ \frac{\sqrt{3}}{2}$$

$$Tan\ 60° \ = \ \sqrt{3} \qquad Tan\ 30° \ = \ \frac{1}{\sqrt{3}} \qquad Tan\ 45° \ = \ 1$$

$$\pi = 180° \qquad \frac{\pi}{3} = 60° \qquad \frac{\pi}{6} = 30° \qquad \frac{\pi}{4} = 45°$$

Sums of different Sines and Cosines

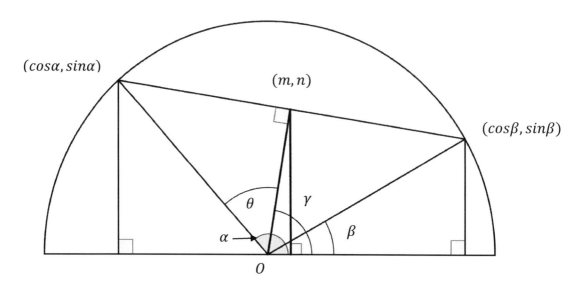

Circle radius 1

A line is drawn from centre which bisects chord at (m, n) making

$$\theta = \frac{\alpha - \beta}{2} \quad \therefore \quad \gamma = \frac{\alpha + \beta}{2}$$

Let perpendicular bisector to (m, n) be h in length

$$\cos\theta = \frac{h}{1} \quad \sin\gamma = \frac{n}{h} \quad \therefore n = \cos\theta \sin\gamma$$

$$\cos\gamma = \frac{m}{h} \quad \therefore m = \cos\theta \cos\gamma$$

$$\textit{from midpoint,} \quad n = \frac{\sin\alpha + \sin\beta}{2} \quad \textit{and} \quad m = \frac{\cos\alpha + \cos\beta}{2}$$

$$\therefore \ \sin\alpha + \sin\beta = 2\cos\left(\frac{\alpha - \beta}{2}\right)\sin\left(\frac{\alpha + \beta}{2}\right)$$

$$\cos\alpha + \cos\beta = 2\cos\left(\frac{\alpha - \beta}{2}\right)\cos\left(\frac{\alpha + \beta}{2}\right)$$

Differences of Sines and Cosines

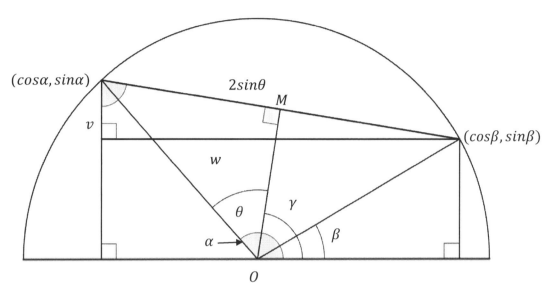

Circle radius 1

A line is drawn from centre which bisects chord at M making

$$\theta = \frac{\alpha - \beta}{2} \quad \therefore \quad \gamma = \frac{\alpha + \beta}{2}$$

a horizontal line is drawn at height sinβ to make right angle triangle of sides $v, w,$ and $2\sin\left(\frac{\alpha-\beta}{2}\right),$ and angle marked is γ

$$\cos\gamma = \frac{v}{2\sin\theta} \quad and \quad \sin\gamma = \frac{w}{2\sin\theta}$$

$$v = \sin\alpha - \sin\beta \quad and \quad w = \cos\beta - \cos\alpha$$

$$\therefore \; \boldsymbol{\sin\alpha - \sin\beta \; = \; 2\sin\left(\frac{\alpha - \beta}{2}\right)\cos\left(\frac{\alpha + \beta}{2}\right)}$$

$$\boldsymbol{\cos\beta - \cos\alpha \; = \; 2\sin\left(\frac{\alpha - \beta}{2}\right)\sin\left(\frac{\alpha + \beta}{2}\right)}$$

Adding or Subtracting Different Sines

$$sin(A + B) = sinA \, cosB + cosA \, sinB$$

$$sin(A - B) = sinA \, cosB - cosA \, sinB$$

Add 2 identities

$$sin(A + B) + sin(A - B) = 2sinAcosB$$

$$let \; A + B = P \; and \; A - B = Q$$

$$\therefore \; A = \frac{P + Q}{2} \quad and \; B = \frac{P - Q}{2}$$

$$\therefore \quad sin \, P + sin \, Q = 2 \, sin \left(\frac{P + Q}{2}\right) cos \left(\frac{P - Q}{2}\right)$$

And similarly

$$sin \, P - sin \, Q = 2 \, cos \left(\frac{P + Q}{2}\right) sin \left(\frac{P - Q}{2}\right)$$

Adding or Subtracting Different Cosines

$$\cos(A + B) = cosA\ cosB - sinA\ sinB$$

$$\cos(A - B) = sinA\ sinB + cosA\ cosB$$

Add 2 identities

$$\cos(A + B) + \cos(A - B) = 2cosAcosB$$

$$let\ A + B = P\ and\ A - B = Q$$

$$\therefore\ A = \frac{P + Q}{2}\ \ and\ B = \frac{P - Q}{2}$$

$$\therefore\ \boldsymbol{cos\,P + cos\,Q = 2\,cos\left(\frac{P + Q}{2}\right) cos\left(\frac{P - Q}{2}\right)}$$

And similarly

$$\boldsymbol{cos\,P - cos\,Q = 2\,cos\left(\frac{P + Q}{2}\right) cos\left(\frac{P - Q}{2}\right)}$$

Maths Quotes

Do not worry about your difficulties in Mathematics. I can assure you mine are still greater. ~ Albert Einstein

Pure mathematics is, in its way, the poetry of logical ideas. ~Albert Einstein

If I were again beginning my studies, I would follow the advice of Plato and start with mathematics. ~Galileo Galilei

"Space is big. You just won't believe how vastly, hugely, mind-bogglingly big it is. I mean, you may think it's a long way down the road to the chemist's, but that's just peanuts to space." ~Douglas Adams

The essence of mathematics is not to make simple things complicated, but to make complicated things simple. ~ S.Gudder

"We should make things as simple as possible, but not simpler." Albert Einstein

Physics is mathematical not because we know so much about the physical world, but because we know so little; it is only its mathematical properties that we can discover. ~ Bertrand Russell

"I refuse to answer that question on the grounds that I don't know the answer" ~Douglas Adams

"The Answer to the Great Question... Of Life, the Universe and Everything... Is... Forty-two,' said Deep Thought, with infinite majesty and calm." ~ Douglas Adams

Glossary

acute ~ sharp or pointed. An acute angle is one which is less than 90°, from the Latin *acus(needle)*

adjoining ~ to be joined to, from the Latin *ad(to) jungere(join)*

algebra ~ using letters or symbols for numbers or quantities, from the Arabic *al-jebr* from the book "Al-jebr w'al-mugabalah" by Abu Ja'far Ben Musa (825 AD) meaning "the reunion of broken parts"

angle ~ space between two lines or surfaces that meet, from the Latin *angulus(corner)*

arc ~ a curve forming part of the circumference of a circle, Latin *arcus(bow, curve)*

arithmetic sequence or progression~ is a sequence of numbers such that the difference between the consecutive terms is constant, from the Greek *arithmetike(to count)*

axis ~ a fixed reference line for use with coordinates, from the Latin *axis(axle,pivot)*

bisect ~ to cut or divide into two equal parts, from the Latin *bi(two) secare(to cut)*

calculus ~ is the study of 'rates of change'. It was developed in C17 independently by Leibniz and Newton. **Differential calculus** determines the rate of change of a quantity. **Integral calculus** finds the quantity where the rate of change is known. Original methods were based on summation of infinitesimals, from Latin *calx(stone)ulus(ule) pebble/small stone used for counting*

chain rule ~ a rule that finds the derivative of a composite function

chord ~ A line segment joining two points on a curve, from the Greek *chorde(gut or string)*

circle ~ 2D shape which is the same distance in all directions from its centre),Latin *circulus(small ring)*

circumference ~ the closed curve(perimeter) of a circle, from the Latin *circum(around) ferre(carry)*

composite function ~ the result of combining two functions, Latin *compositus(to have put together)*

cone ~ 3D shape with circle as base, from which all sides meet at an apex, Greek *konos(cone)*

congruent ~ shapes which are the 'same' ie. identical except perhaps for orientation. In triangles shared properties for congruence: 3 sides, 2 angles and 1 side, 2 sides with enclosed angle or a right angle and 2 sides, from the Latin *congruere(coming together,agree)*

co-ordinates ~ numbers within a graphical system which together determine the position of a point, ie. values on the x and y axes, from the Latin *coordinare(to set in order,arrange)*

cyclic quadrilateral ~ quadrilateral inscribed within a circle

cylinder ~ a circular prism, from the Greek, *kylindros(cylinder,roller)*

degree (angles) ~ a unit measurement of angle, 360 degrees making a full circle, dating back to ancient Babylonians, (who used base 60) being divisible by 2,3,4,5,6,10,12,15, Latin *degradus(a step)*

denominator ~ the divisor component of a fraction

derivative ~ the function representing the rate of change of an original function

diagram ~ that which is marked out, from the Greek *dia(through) gram(written or drawn)*

diameter ~ chord of a circle which goes through the centre, Greek *dia(through) metros (measure)*

differentiation ~ the process of determining the derivative of a function,

displacement (transformations) ~ the distance that a function is displaced, ie. in x or y direction.

dot product $a \cdot b$ ~ scalar product of two vectors

equation ~ a statement of equality of two expressions or quantities, using an equal sign between them

equilateral triangle ~ a triangle with three equal sides

exponent ~ power or index

exponential curve ~ of the type $y = a^x$, or commonly e^x, exponential(rapid) growth and decay

f(x) ~ function of x, **f '(x)**~ derivative of f(x), **f "(x)**~ derivative f '(x)

factor ~ A number or algebraic expression by which another is exactly divisible, Latin *factor(dooer)*

formula ~ A mathematical relationship or rule expressed in symbols, from Latin *forma(shape,mould)*

function ~ relationship for which each input has a single output

geometry ~ properties and relationships of size, shape and space, Greek *geo(earth)metria(measure)*

geometric progression ~ sequence of numbers where the next term is the product of the previous term and the same constant, the common ratio r. First term a, nth term ar^{n-1}

golden ratio ~ special ratio of two quantities, the ratio of sum to the larger is equal to the ratio of the larger to the smaller quantity. Also known as divine proportion, this ratio appears in nature and is linked to the Fibonacci sequence. It also appears in architecture including the Egyptian pyramids.

gradient ~ The steepness of a slope expressed as the ratio of vertical change to horizontal change.

hemi sphere ~ one half of a sphere, from the Greek *hemi(half) sphaira(sphere)*

hypotenuse ~ the longest side of a right triangle opposite the right angle, Greek *hypoteinousa (stretching under)*

increment ~ small value of change in a quantity, represented by δ

indices ~ plural of index, exponent or power

infinity ~ an unbounded limit ∞ , from the Latin *infinitatem (boundlessness,endlessness)*

integer ~ whole number, positive or negative and zero, from the Latin *integer(whole)*

inscribed ~ a geometric shape within another, which touches at all possible boundaries

integration ~ the inverse process of differentiation, from the Latin *integratus(make whole)*

intercept ~ intersection of one line with another

isosceles triangle ~ triangle with two sides equal, from the Greek *isoskeles(with equal sides)*

LHS ~ left hand side   **RHS** ~ right hand side

logarithms ~ The exponent or power to which a base number can be raised so that the result is a specified number, coined by John Napier (1550-1617), from the Greek *logos(ratio) arithmos(number)*

mass ~ the amount of matter in a particle or object, Latin *massa (dough,lump,mass)*

modulus |x| ~ the magnitude or absolute value of x (non negative)

natural logarithm ~ logarithm with base e, $log_e x = \ln(x)$,

negative ~ less than zero, with a minus sign before, in use since 1706

nth term ~ an expression that gives the term in nth position in the sequence

numerator ~ the top half of a fraction, the portion which is to be divided

parallelogram ~ shape bounded by parallel lines, from Greek *parallelos(parallel, beside each other)*

pentagram ~ five pointed star, from the Greek *pentagrammon(having five lines)*

perpendicular ~ geometric objects being at right angles with each other,Latin *perpendicularis(vertical)*

pi (π) ~ the ratio of the circumference to its diameter, named by Euler in 1737

polygon - shape made from a number of straight sides, from the Greek *polygonos(many angled)* with 5 sides pentagon, 6 hexagon, 7 heptagon, 8 octagon, 9 nonagon, 10 decagon

polynomial ~ an expression including constants and variables and non negative exponent terms, from the Greek *poly(many) nomial(terms)*

positive ~ greater than zero, with an assumed plus sign before

prism ~ 3D shape whose cross section is the same all along its length, from the Greek *prisma (something sawed)*

product ~ the result of multiplication

progression ~ sequence, arithmetic or geometric

proportion ~ A part, share, or number considered in comparative relation to a whole

quadratic equation ~ equation with variable to maximum power 2(squared),Latin *quadratum(square)*

quadratic roots ~ value of variable that makes the function equal zero, where it intercepts the x axis

quadrilateral ~ a polygon with four sides, from the Latin *quadri(four)lateris(side)*

radian ~ unit measurement of angle, which is a pure number, ratio of arc length to radius. The radian was defined and named by James Thomson in 1873, brother of Lord Kelvin

radius ~ a line segment from the centre to any point on circle or sphere, from the Latin *radius (ray of light, spoke of wheel)* first used geometrically 1610

ratio ~ relationship of two or more quantities in proportion with each other, which can be expressed as a number of shares or parts of a whole to (:) each other, from the Latin *ratio(reckoning)*

rational and irrational ~ a rational number can be expressed as a ratio of two integers, an irrational number cannot

re-arrange (algebra) ~ to manipulate algebra equations or formulas by doing the inverse operation to both sides of the equal sign to gain a value/new identity for a variable

reciprocal ~ the multiplicative inverse ie. 1 ÷ by the original number, Latin *reciproco(to reverse)*

rectangle ~ quadrilateral with four equal right angles, from Latin *rectus angulus(upright angle)*

regular polygon ~ polygon with equal sides and equal angles

right angle ~ angle of 90° which makes a perpendicular, from Latin *rectus angulus(upright angle)*

scalar (vectors) ~ a quantity with only magnitude not direction

sector ~ section of a circle between two radii and the circumference

segment ~ part of a circle between a chord and the circumference, Latin *segmentum(cut into parts)*

semi circle ~ half a circle, from the Latin *semi(half)*

sequence ~ a set of terms with a rule applied to the members, from the Latin *sequens(following)*

series/sums ~ a series is a sequence with all the terms added, a sum is the total sum of all the terms

similarity ~ a property of geometric shapes of sameness, so that applying a scale factor would make them congruent

sphere ~ 3D shape which is the same distance in all directions from its centre, Latin *sphaera(sphere)*

square ~ a regular quadrilateral, with equal sides and angles, from Latin *exquadrare(to make square)*

subtended angle ~ the angle formed by two rays from an object or two points, to an external vertex, from the Latin *sub(under)tendere(stretch)*

symbols ~ therefore $\therefore$, equivalent to $\equiv$, tends to $\rightarrow$,less than $<$, greater than $>$,integral $\int dx$

symmetry ~ reflection symmetry where an object is identical when reflected in an axis (line of symmetry), from the Greek *symmetria(agreement in dimensions)*

tangent ~ a line which touches a curve at one point, from the Latin *tangere(to touch)*

theorem ~ A general proposition not self-evident but proved by a chain of reasoning; a truth established by means of accepted truths. From the Greek *theorema(proposition,speculate)*

trapezium ~ a quadrilateral with one pair of parallel sides, from the Greek *trapeza(table)*

triangle ~ a polygon with three sides, from the Latin *triangulum(three cornered/ angled)*

triangular number ~ the number which is the sum of natural integers from one to n which can be represented by a triangle

trig, trigonometry ~ study of relationships of sides and angles of triangles, Greek *tri(three) gonia(angle) metron(measure)*

turning point ~ also known as a stationary point. A point on the graph of a function where the tangent has gradient zero

vector ~ A quantity with both direction and magnitude represented by an arrow, which gives direction, and length which is proportional to the magnitude, from the Latin *vector(one who carries or conveys)*

Notes

p6-7   negative and positive

for multiplying/dividing $-\ -\ \rightarrow\ +$

or double sign in front of number $+\ +\ \rightarrow\ +$

$-\ +\ \rightarrow\ -$

$+\ -\ \rightarrow\ -$

p58   Sine Rule can also be shown by $\dfrac{Sin\,A}{a} = \dfrac{Sin\,B}{b} = \dfrac{Sin\,C}{c}$

p59   Cosine Rule also

$$b^2 = c^2 + a^2 - 2ac\ CosB, \quad and \quad Cos\,A = \frac{b^2 + c^2 - a^2}{2bc}$$

$$c^2 = b^2 + a^2 - 2ba\ Cos\,C$$

p76-77 Versluys can also be shown by a tessellation of squares side a and b with a square of side c suitably inscribed

p85   Vectors

parallel vectors of the same magnitude are equal $a \nearrow\ a \nearrow$

p102   Sum of arithmetic progression

Often the proof for this uses 2 sums of a .... a+(n+1)d which provides n terms of 2a+(n−1)d which covers a solo middle term, However this proof is more vivid, and a 'middle' term would simply be $a + \left(\frac{n-1}{2}\right) d$ which fits since the sum is also $n(a + \frac{(n-1)d}{2})$

p118   Integration as area and volume

because the area/volume is found from 0, find $\int_0^4 y\ dx\ and\ subtract\ \int_0^1 y\ dx$

27966607R00083

Printed in Great Britain
by Amazon